What Dogs Know

Juliane Bräuer • Juliane Kaminski

What Dogs Know

 Springer

Juliane Bräuer
Max Planck Institute for the Science of
Human History
Jena, Thüringen, Germany

Juliane Kaminski
Department of Psychology
University of Portsmouth
Portsmouth, UK

Translated by
Neil Solomon

Translation from the German language 2nd edition: *Was Hunde wissen* by Juliane Bräuer und Juliane Kaminski, © Springer-Verlag GmbH Deutschland, ein Teil von Springer Nature 2020, Published by Springer-Verlag Berlin Heidelberg. All Rights Reserved.

1st edition: © Franckh Kosmos Verlag 2011

ISBN 978-3-030-89532-7   ISBN 978-3-030-89533-4   (eBook)
https://doi.org/10.1007/978-3-030-89533-4

Cover photo: © Juliane Bräuer
Illustrations: Nora Tippmann. With 53 photographs by Viviane Venzke/Kosmos

This Springer imprint is published by the registered company Springer Nature Switzerland AG.
The registered company address is: Gewerbestrasse 11, 6330 Cham, Switzerland

Acknowledgements

We would like to take this opportunity to thank all those who have helped us to produce this book. And we would like to apologise to all those who had to put up with us being busy with the book for weeks on end, especially our families. Special thanks go to Matthias Braun, Carmen Bräuer, and Angela-Maria Chira for proof-reading the book.

Contents

Contents

About the Authors

Juliane Kaminski and **Juliane Bräuer** have been studying the cognitive abilities of dogs for 20 years. The authors both studied biology, then completed their doctorates at the Max Planck Institute for Evolutionary Anthropology in Leipzig, and conducted numerous behavioural studies.

Juliane Kaminski is an assistant professor at the University of Portsmouth, where she heads the Dog Cognition Centre. She is interested in communication between humans and dogs and is co-editor of the book, *The Social Dog: Behaviour and Cognition.*

Juliane Bräuer heads *DogStudies* at the Max Planck Institute for the Science of Human History in Jena. She is interested in what abilities dogs have developed over the course of domestication. She is author of the book, *Klüger als wir denken: Wozu Tiere fähig sind.*

1

Why Dogs?

In June 2004, the dog world was abuzz. Rico, a 9-year-old Border collie, had proved under scientific conditions that he could distinguish 200 different objects by name. And not only that, he could also learn new names for toys in a way that had previously only been demonstrated in children. The press spoke of the 'Einstein of the dog world', and a major German newspaper even went so far as to say: 'Finally proven: dogs understand German'. It seemed as if one of the big questions had been solved! Dogs understand us. They understand every word! But is that really so?

What do dogs understand about us humans? Do they understand what we are trying to communicate to them? What do they understand about the world they live in? These are the questions this book is about. It is about *cognition*, specifically, the *cognitive capabilities* of domestic dogs. We are interested in how dogs perceive their environment and what knowledge they can gain about it. On the one hand, it is about what dogs understand about their inanimate environment. For example, do dogs understand simple relationships such as gravity? Do they know that objects that are dropped always land on the earth? On the other hand, it involves the question of what dogs understand about their animate—i.e., social—environment. What do dogs understand about us humans? How do we communicate with them and they with us? In the course of this book you will come to see that this is clearly where our pet's great talent lies. Would you have thought, for example, that your dog can tell whether your eyes are open or not? Or that objects become exciting for it solely because *you* have touched them?

© The Author(s), under exclusive license to Springer Nature Switzerland AG 2021
J. Bräuer, J. Kaminski, *What Dogs Know*, https://doi.org/10.1007/978-3-030-89533-4_1

Why Testing Dogs

But why are we dealing with the cognition of domestic dogs? When we think of cognition, certainly the great apes are the first thing that comes to mind, because they are so closely related to us humans. Or, at the very least, one thinks of animals that have to make it in the wild. Like wolves, for example. But one doesn't tend to think about house pets.

In fact, even in science, dogs have long been regarded somewhat pejoratively as domesticated, as 'imperfect' wolves. They have lost many of the abilities of their wild ancestors. For example, they have smaller brains, and they are also thought to have worse hearing and smell than their wild relatives. Many scientists have rejected research involving dogs because dogs supposedly live in an 'artificial' environment. Results of such experiments would not reflect 'true nature'.

Recently, however, it has become accepted that dogs do not live in an *unnatural* environment, but simply in a *different* environment than wolves (Fig. 1.1). The human environment is the natural environment for dogs. This means that *different* skills are required here than in the wild.

Since dogs have lived together with humans for a very long time, many scientists increasingly see our four-legged friends as an interesting model. Because during this long coexistence they may have developed special abilities to adapt to life with humans. Since the turn of the millennium, this rethinking in the way that dogs are viewed has given rise to numerous new studies.

Fig. 1.1 Much has changed in the lives of dogs compared to those of the wolves

About This Book

A lot is written and claimed about dogs. There are a lot of books in which the behaviour of the dog is interpreted and sometimes humanised. In this book we would like to present, in a vivid manner, the latest results from science. All of these findings are based on studies conducted by scientists the world over. Our aim is to explain how such studies are conducted, what their results mean, and how they are interpreted. Thus, what you read here is not our personal assessment or opinion. Of course, this still does not make the book completely objective. We have had to select from many studies, and we may have overlooked some. Others have been left out since we found them unimportant or unconvincing because, in our opinion, they were poorly conducted methodologically. Nevertheless, we have made every effort to give you a very broad overview of what we know about dogs to date.

Parts of this book were published years ago—but we have now reviewed and updated it in its entirety. The many new studies have changed science's image of dogs a bit in the last 10 years. The focus of research has shifted and new methods have been used, such as magnetic resonance imaging and physiological measurements. This has resulted in completely new chapters in this book—such as the one on cooperation between humans and dogs—whereas others have been completely rewritten or shortened. All in all, our book aims to give you an up-to-date and comprehensive picture of research on dog cognition.

What Do We Need Studies for?

Time and again we are asked by dog owners: why do we need these tests in the first place? Every owner should be in the best position to know about his or her pet. After all, the owner observes it for most of the day. And it is true that such observations always provide good ideas for new tests. The problem lies in how the dogs' behaviour is interpreted. There are dog owners who unhesitatingly attribute a high level of understanding to everything their dog does. They measure each and every behaviour with human yardsticks. This sometimes goes so far that no differences between humans and dogs are even perceived.

We ourselves are dog owners and know from our own experience how quickly an explanation can be found that puts our own dog in the best possible light. However, there are always several possible interpretations for any given behaviour. The tests serve to find the right explanation for a specific behaviour and to exclude alternative interpretations.

Let's take an example that many of you are probably familiar with. You want to take your dog to the park. Before you can go, however, you need the toy that makes your dog happy outdoors. You look around, but you can't find it. Then you see that your dog is standing in the corner of the room by the wall unit and looking back and forth between you and it. Because you do not react immediately, it becomes even more emphatic and finally jumps up on the wall unit and runs back and forth. Finally, you understand what your dog is trying to 'tell' you. The toy is in the wall unit. You open it up and, lo and behold, there it is.

How can you explain your dog's conspicuous behaviour? Had your dog known exactly what you were looking for and sought to inform you about where the toy was? People communicate in this way. If this communication had taken place between two people, it would certainly have been meant in this way. One person recognises the need of the other and responds by indicating what is needed, helping in this way. It is possible that your dog is behaving in exactly the same way in the situation described above. However, there are other possible explanations for your dog's behaviour. Perhaps your dog knows that the walk is about to start. It notices this, for example, because you have put on your shoes or grabbed the lead.

Your dog may then simply automatically point to the toy because it has a selfish interest in having this toy taken along. Thus, it is not concerned with helping you find it through its communication. It is only concerned with its need to get the toy. Since it cannot get the toy on its own, it needs you as a kind of 'tool' to get the toy for it. This leads to two possible explanations. The first hypothesis would be that dogs communicate to help you search because they recognise your need. The second hypothesis would be that they act solely out of selfish motives to gain an advantage for themselves.

So how can we find out what your dog's behaviour really means? This is the moment at which a test can help. A test is necessary to really know what is behind a dog's behaviour. So-called *controlled* conditions are created. This means that we confront our test dogs with a certain predefined situation. We make sure that the procedure is always the same. Only certain parts of this sequence are changed in a deliberate way. This is what we call the variable conditions. Now we compare the dogs' behaviour under these conditions. This allows us to isolate the reason why the dog behaved the way it did.

It is important in these tests that the experimenter tries to be unbiased. Because you often see what you want to see. In addition, if possible, the behaviour of the dogs in the trials should be recorded with a video camera.

On the one hand, this allows for the material to be evaluated at leisure. On the other hand, this also allows it to then be examined by other, completely unbiased persons. In the best case, by a person who does not even know the question the study seeks to answer. This person should evaluate the behaviour of the dogs according to clear and unambiguously defined criteria. Only if the unbiased observer agrees with the researcher directing the study in his or her evaluation of the behaviour, are the results of the study credible.

But the evaluation of the video does not mark the end of our experiment. Because now comes the statistical evaluation. Rarely are the results completely unequivocal on first sight. Rarely will *all* the dogs in one group behave differently from *all* the dogs in the other group. That is why we need statistical tests. They answer the question: Can our results be explained *by chance*? If so, would the dogs have behaved completely randomly and independently of which test situation they were in? Or are our results *not random*? That would mean the dogs' behaviour really had something to do with the different conditions they were in. If our results are not based on chance, then they are meaningful.

But how do we interpret the behaviour of our four-legged friends? If, for instance, a dog does *not* 'pass' a test, there may be several possible reasons for this. Perhaps the test design was not sensible. Maybe there was something that distracted the dog? Was it motivated? Is it capable, in purely physical terms, of solving this problem? If you have eliminated such flaws and the dog still fails, it is probably the dog's lack of understanding that keeps it from solving the problem. And if the dogs can solve the problem? Does that mean they understand as much as we do? Or do they perhaps use a completely different strategy and arrive at the same result? Did the animals perhaps learn the solution in the course of the experiment? We have to check this too, because we are more interested in the dogs' spontaneous behaviour and less in their learned behaviour. These are all questions that need to be answered before a statement can be made about the dogs' abilities.

Fun for Dogs or Animal Experiment

Dog owners tell us what they have gone through with their pets. These observations, as well as our own, provide the ideas for studies. However, the overriding principle in this kind of work with dogs is that everything is done on a voluntary basis. It should be fun for everyone involved. Only then can the dog attain its best results. The situations created for the purpose of the study are usually very playful.

In most experiments, a problem of some kind must be solved. The stimulus is either food or toys (Fig. 1.2). For example, a ball has to be found. Or a lever has to be pressed to get food. When the dogs have solved the problem, they are allowed to eat the food. Or they play a round. These motivational phases in-between help to keep the dog in a good mood.

For this type of study, it is important that the dogs do not lose their motivation. This could falsify the results. In such a case, the animal might be able to solve the problem, but it 'no longer feels' like doing anything. So it is in the interest of all involved that the dogs never stop having 'fun' at their 'work'. Most of the dogs seem to like our tests very much. No matter whether in Leipzig, Jena, or Portsmouth…

Dog owners are always telling us that their pets turn off on their own toward the dog testing rooms on their way to the park. So it seems that a new test is better than a familiar walk. In the testing rooms the dogs get to play, search for food, and—if desired—make contact with their own kind. And above all, there is a human being there who is only concerned with them.

Depending on the question asked in the test, the owners are sometimes part of the study and sometimes not. To keep the general conditions of our tests as constant as possible, many studies are carried out without the dogs' owners. This is because dogs are extremely sensitive to the slightest cues from their owners. Thus, it is our aim to keep owners from giving their animals these small forms of assistance. An assistance that can even take place unconsciously. After all, it is almost impossible to suppress the desire to have one's

Fig. 1.2 Both are good motivators: toys and food

own dog master the task at hand. But what interests us most is how the dogs can solve the problem on their own!

However, there are studies that are dependent on the participation of the owner. For example, if you want to know how a dog and its owner communicate with each other or whether dogs prefer to assist people they know over those that they don't. Then, of course, the owner is very important for the test. Sometimes he or she is given very precise instructions on how to behave in the test—depending on the conditions in the particular study. In other cases, the owner is allowed to move freely with his or her dog and behave as he or she would if he or she were unobserved. Here the free interaction between dog and owner is recorded and analysed.

Some of the experiments described in this book can be easily replicated at home. You will therefore always be encouraged to try one or the other with your dog. Such a test may provide you with more than a few new insights into your own pet. It may also ensure that your four-legged friend has an interesting afternoon.

The Comparison

You might be surprised to discover that this dog book is not only about dogs. It often includes work with children, monkeys, and even goats (Fig. 1.3). It is not that dogs alone do not provide enough to talk about. It is because this comparative approach plays an important part in cognitive research. This raises the question about the evolution (i.e., the development) of cognitive abilities.

There is no doubt that humans have developed a very high level of cognitive ability. This applies both to their acute understanding of the inanimate world (quantities, colours, shapes, etc.) and to their sophisticated understanding of the social environment in which they operate. We know in some

Fig. 1.3 Making comparisons between species is important for understanding how cognitive abilities have developed in the course of evolution

situations what others can or cannot see. We understand what they are trying to tell us. In certain cases, we can even discern the aims and intentions of the person (or animal) opposite us. We make use of this knowledge on a daily basis. This means that we humans are able to perceive others as autonomous agents with their own thoughts, feelings, and ideas. Comparative cognition research asks the question whether other species are also able to do this. Studies with apes seem to be the obvious choice in this context. Great apes are the closest living relatives of humans. So if we were to find comparable abilities, it would be here.

But dogs? How could dogs fit into such investigations, apart from the fact that, as a dog lover, one naturally likes to work with these animals? There are two things that make dogs interesting. First, they are mammals. They belong to the same systematic group as humans and apes. However, they are more distantly related to humans than apes are. This makes them interesting for comparative studies. If you compare different animal species with each other, you can see how widespread certain abilities are in the animal kingdom. This comparison is not only important with species closely related to us, but also with those more distantly related.

We have already mentioned the second reason why dogs make an interesting comparison. It has to do with their long and intensive coexistence with humans. It is assumed that dogs have developed special abilities in this relationship. When we hear, for example, that the Border collie Rico learns names of objects by matching, we may not be that surprised. However, when we learn that this ability has otherwise only been shown to exist in humans, a completely different picture emerges. This book would like to draw your attention to these distinctive features of our dogs' abilities.

When looking at the behaviour of dogs, the comparison with their closest relative, the wolf, is of particular interest. Especially when the question arises about the extent of influence that the process of domestication had on dogs' abilities. An entire chapter of this book is devoted to this comparison. Similarities between dogs and wolves tell us something about the abilities of their species in general. Differences show what has changed in the long period of dogs' domestication.

In comparison with other species, we see that dogs can do some things particularly well and other things not as well. This raises the question of *why* this is so. It is related to the skills that their environment calls for. This is where their 'talents' lie. The studies are basically designed to require skills that are also relevant to dogs. Accordingly, we don't test whether dogs can talk. Or whether they can divide a number by 17. Instead, we try to give test dogs tasks that are relevant to them in their everyday environments.

We are, of course, often asked the question: How smart are dogs? If you compare their abilities with those of small children, how smart are they? As smart as 2-year-old children or as smart as 5-year-olds? Clearly, it is not that easy to answer this question. It all depends on the task. As you will see when reading this book, there are areas in which dogs perform as well as 3-year-old children. In other areas, they can't even hold a candle to a 10-month-old child. And in one area they probably do even better than humans—due to their extremely sensitive noses. Dogs may well understand more about the world of smell—and what is connected with it—than we humans do. But we don't know much about that yet.

You may notice that we never speak of 'intelligence' in the following pages. This is not to say that dogs are not intelligent. But we prefer to speak of cognitive abilities. Because this expression allows for a differentiated view. Dogs are not intelligent or stupid. It is just that they can do some things and not others.

2

How Wolves Became Dogs

A scene at the dog run: A golden retriever is tussling with a German shepherd. A St. Bernard stands calmly nearby. It weighs in at a hefty 85 kg. Its head alone is bigger than the entire dwarf dachshund that lifts its leg a few metres away. In between, graceful greyhounds flit back and forth, chasing and catching each other. A Labrador never tires of fetching, again and again, the ball thrown by its owner. A Puli joins in. Its face can only be intuited under its long coat.

It is hard to believe that all these different animals are closely related and are supposed to share the same progenitor: the wolf. For a long time, this resulted in speculation whether other canids, such as golden jackals or coyotes, could be the closest relatives of dogs. However, it has also been known for a long time that dogs very rarely—if ever—mate with coyotes or jackals, neither in the wild nor in crossbreeding experiments. They do, however, mate with wolves. All recent studies also speak against coyotes or jackals playing a role as dog ancestors. Geneticists, morphologists, and behavioural scientists agree: wolves are the closest living relatives of our present-day domestic dogs (Fig. 2.1). There is some debate among anatomists as to whether dogs represent a species of their own (*Canis familiaris*) or a subspecies of wolves (*Canis lupus familiaris*). But that need not concern us here.

When It Began

In the following, it is important for us to distinguish between the progenitor wolf and the wolf of today. For today's wolf is not the direct ancestor of our dog, but its cousin. Thus, both dogs and present-day wolves are descended

J. Bräuer, J. Kaminski, *What Dogs Know*, https://doi.org/10.1007/978-3-030-89533-4_2

Fig. 2.1 Clearly proven: dogs and wolves share a common ancestor

from a common ancestor, just as two cousins have grandparents in common. When we look at the differences between dogs and wolves, we have to bear this in mind: we cannot compare the dogs in our studies today with the progenitor, the 'primeval wolf', but only with its cousin, who lives today. And, of course, this cousin has also evolved over the last 40,000 years. The wolf has, for instance, been subject to intense hunting in many parts of the world. So you can imagine that only the most reclusive wolves have survived and reproduced. Among dogs, on the other hand, those that were particularly good at getting along with people prevailed.

As far as dogs are concerned, everything suggests that they all—from the miniature pinscher to the Great Dane—are descended from the 'primeval wolf'. It is also certain that dogs are the oldest domesticated animals. They have lived with us much longer than, for example, goats, which were the next animal to be domesticated. But when exactly did it happen, when did wolves become dogs? When did they develop floppy ears and start barking? It is not so easy to answer this question from the genetic material. It is possible to look at how closely dogs and wolves are related to each other. But we can only guess when the split between the species took place. Biologists still do not agree here, because different genetic methods have led to different results. In the meantime, agreement has been reached that domestication of the dog began about 35,000–40,000 years ago.

Archaeologists can confirm this. Since the Ice Age, bones of wolves have been found close to those of early humans. This may be because both species were widespread at the time. Perhaps the relationship between them had already begun then. But we can only speculate about that. Scientists agree on the oldest bone find that is clearly from a dog. It is a jawbone from an excavation near Oberkassel in Germany. The bone is about 14,000 years old. A whole series of findings from the period thereafter confirms that dogs had already existed all over the world by that time.

According to this, dogs were domesticated at a time when we humans were not yet settled. They moved around with our ancestors and settled new areas with them, for example, the American continent. When Columbus discovered America over 500 years ago, there was only one domestic animal there that was also known in the Old World: the dog. A comprehensive genetic study has shown that the Native Americans did not actually domesticate dogs themselves. The first migrants to America were already accompanied by dogs when they settled the continent. For it could be proven that the American dogs had Siberian ancestors and were not descendants of American wolves.

This study on the origin of American dogs had another surprising result: evidently, the dogs in North America shared the sad fate of their owners. Just like the indigenous people of North America, their dogs were severely depleted in population. As a result, hardly any genetic traces of them can be found in today's dog population in the USA and Canada.

Dogs and humans colonised the earth together—and, all in all, did so relatively quickly. However, they encountered difficulties in getting south of the 20th parallel. For dogs did not arrive in the tropics until 5000–3000 years ago. Earlier authors assumed that dogs had not spread there because they were not that useful. It was initially assumed that they were not much help in hunting in the tropical rain forest because of the density of the vegetation there. It is now assumed, however, that diseases were the reason dogs did not make it across the 20th parallel. It apparently took thousands of years before dogs adapted to the pathogens of distemper, leishmaniasis, and sleeping sickness. But once they had overcome this hurdle, they became native to the south— just like everywhere else where humans are found.

Different types of dogs, which can be distinguished by appearance and usefulness, can first be found 3000–4000 years ago. There were greyhound types that had apparently been used for hunting. Others of large and powerful stature were perhaps used as watchdogs. Systematic breeding, as we know it today, took place for the first time in the Roman Empire. Since then, new breeds of dog have been created time and again, especially in times of cultural prosperity. In times of crisis, these breeds mixed with others and disappeared. It is not uncommon for them to reappear later in a similar form. Today we can count about 400–450 breeds, which differ considerably not only in size and appearance, but also in their individual behaviour and temperament.

How It Began

Humans and dogs have thus spent quite a long time together. But how did they first draw closer to one another? How did man come to have a dog? Today we can only speculate about this.

For a long time, it was assumed that the initiative for living together came from humans. It was imagined that the prehistoric hunter killed a she-wolf and took in the puppies. Women may have also played a special role here. With their large round head, short round snout and clumsy movements, the puppies made an endearing impression and may have awakened maternal instincts. After all, a puppy's gaze is practically irresistible; we'll come back to that below. Perhaps it all started like this: a woman had just lost her newborn and began to care for a helpless wolf pup. Perhaps she even breastfed it. Why do we consider such a sentimental story likely? We know that only very young wolf pups get used to humans. Because at a very young age, pups are still dependent on milk. At that time, however, there were no other domestic animals. And that in turn means that they must have been fed on human milk. You may find it quite far-fetched that a female human would breastfeed a baby dog. However, this was common in many more cultures than you might think: especially in Southeast Asia, Australia, Tasmania, Oceania, and throughout the Americas. It is therefore likely that the nursing of puppies contributed to domestication.

Wolf pups were thus raised and tamed by humans. But what happened if the soon-to-be adult wolf did not behave submissively? People certainly did not want to have a housemate who could not be controlled. If the wolf was aggressive or indeed attacked people, it was certainly killed or chased away. If, despite human care, it was shy and skittish, it would probably return to the wild of its own accord. So only those animals remained in the village that were unafraid and submitted to humans without further ado. There they could then reproduce.

There is another imaginable scenario of how humans and dogs began to live together. Most scientists now consider this one to be the more likely. The wolf may have domesticated itself. Let's assume we have a larger group of perhaps 15 wolves. In such a group, they are never all identical, neither in anatomy nor in personality. Let's say there are five relatively tame individuals in our group who have little fear of humans. The rest of the pack is not so brave, but these five have no problem being around humans. Because of their tameness, these animals now enjoy an advantage. For where humans hunt, there will be scraps, which might make good and easily accessible food. Aside from this, there might have been a second advantage. Wolves were better protected from their predators in the vicinity of humans. So it could be that our five wolves carved out a little niche of their own. This advantage, in turn, could have resulted in them producing more offspring of their own. And their offspring may well have had more 'tame' genes than the offspring of the other 10 wolves from the original pack. The longer the 'tame' wolves spent near humans and reproduced there, the less shy they became about getting close to humans. And so, over several generations, a whole new population of so-called 'proto-dogs' emerged. They still looked wolf-like but were less wolf-like in behaviour with each new generation.

We may never know exactly how dogs and humans came together. A mixture of the two scenarios described is certainly also conceivable: Humans and wolves may have come closer together in equal measure from both sides.

Another question is where geographically did the 'primeval wolf' become a dog? Did this happen in many places at the same time, or perhaps only once? This is a question for geneticists who study the genetic material of animals. Some argue that the process of domestication only took place once or twice in eastern Asia. After that, they assume, dogs spread all over the world. Others doubt this and assume that a coming together between the 'primeval wolf' and man must have occurred in several places. Findings differ considerably depending on which genetic material is studied. For example, there is much more genetic diversity among street dogs, which have been neglected in research so far, than among pedigree dogs. But why do geneticists have such a hard time finding a clear answer? It is easy to explain this once you consider that there has obviously always been contact between wolves and dogs over the millennia. Down through history, dogs have mated and exchanged genetic material with their wild relatives again and again. It is this repeated exchange that complicates the research of geneticists today.

Why It Began

So far, we have been looking at how the relationship between humans and dogs came about. Now we want to ask why it came about. Why, for example, was the wolf of all animals domesticated? A decisive point was certainly that both humans and wolves are very social creatures. There are also astonishing similarities between the species in other respects: Both have pronounced communicative abilities. Both form personal bonds with the individual members of their group. Both parents take care of their offspring, as does the whole group. Both play with their children. At the time when humans and wolves met, both lived in similarly structured family groups with social hierarchies. Humans and wolves were obviously well suited to each other, but what benefits did they derive from this relationship? The benefit for the animals—easier access to food—has already been reported.

There is much speculation about the benefits for humans. One important aspect is certainly that dogs helped humans to hunt. There is an interesting theory that tries to explain why dogs were domesticated at the end of the Ice Age. It has to do with our diet. Because humans cannot only consume proteins, they need to cover half of their energy needs with carbohydrates or fats. At times when no plant carbohydrates were available, humans had to resort to animal fats. At the end of the Ice Age, there was a sharp decline in particularly large prey. But smaller animals are more costly to hunt, and you also need more of them to feed the same number of people. In addition, larger animals store more fat than smaller ones do. Logically, it is much more effective to kill a single mammoth than several deer. But because mammoths became extinct after the last Ice Age, humans had to go to much greater lengths to meet their energy needs. And this is where the dogs come in. They may have been a welcome help in this situation. When people began to hunt with bows and arrows, dogs were able to track down and corner fleeing game. However, it is not so easy to imagine how the still very wolf-like dogs voluntarily left the hunted prey to humans.

Perhaps man also first took advantage of their vigilance. He could watch them to know when danger was imminent. Probably they then simply retreated. Maybe the first dogs were already barking. But that would have alerted the predators. And it's unlikely that dogs at this early stage would have risked their lives to defend a human.

They might also have kept the living space clean. They not only ate the human's rubbish, but also served as a nappy substitute. The dogs stayed near the children and ate their 'droppings'. This was common, for example, among the Turkana tribe living in Kenya. Another option was to use the dog as a 'heating pad', as the Australian Aborigines did on cold nights. In other cultures, the skins of slain dogs were used for the same purpose.

Dogs were also used as pack animals. Since our ancestors lived as nomads, the animals may have carried their belongings when a group moved from place to place. This was the practice of some Native American groups before mustangs were tamed. Considering that large dogs can carry 15–20 kg, they would have been a great help during treks.

Perhaps the early dogs were also eaten, as is still not uncommon in some cultures today. This was mainly done as part of special rituals. However, this will hardly have been a reason for domestication, because humans and dogs are food competitors. In principle, they have a similar diet. Thus, it would not have made sense at all to 'fatten up' dogs in order to then consume them.

Plant-eating goats are much better suited for this. Goats followed dogs as the next domestic animal to be domesticated, between 9000 and 12,000 years ago. Presumably, dogs were then used relatively soon thereafter to tend herds of goats. This was certainly not the primary original use of the descendants of wolves.

Thus, we are not able to say with complete certainty how the very first dogs were used. What we do know for sure, however, is that dogs played a crucial role in hunting. Because they can be a great help at this. This is impressively proven by a study from Helsinki, Finland. Here, data on hunting success were analysed. The scientists analysed over 5000 elk hunts. They were particularly interested in the comparison between hunters who had dogs with them and those who did not. They also wanted to know what influence the number of hunters had on the success of the hunt.

The results were clear. With dogs, the hunt was much more successful. Hunters accompanied by dogs killed up to 56% more elk than their dogless colleagues! If the hunters were travelling alone, they needed a certain group size if they wanted to be successful. However, if they had a dog with them, fewer hunters were sufficient for optimal success—regardless whether 9 or 19 hunters hunted together. As soon as more than ten hunters were on the move, several animals proved helpful. With them, a larger area could then be searched for game.

This is one of the first studies to systematically show the great benefit of dogs in hunting large game. You may not find these results especially surprising: hunting dogs are useful. But if you think about what this meant for prehistoric man, things look a little different. Even *a few* hunters with a dog proved to be *highly* successful and killed much more game than without one. Thus, thanks to their domesticated companions, it was easier to ensure the survival of the group.

Whatever brought humans and dogs together, it must have been of great advantage to both sides. It was and remains a successful union. For as we have already seen, dogs, alongside man, spread rapidly over the entire earth. If dogs really were domesticated as early as most geneticists suppose, then there were even Neanderthals around at the time. And perhaps the newly formed human–dog team was so successful in its cooperative efforts that it even contributed to the extinction of the competing Neanderthals. At least that's what an American anthropologist assumes. However, there is still no real evidence for such a far-reaching claim.

In any case, it is a fact that domestication eventually led to dogs living in dependence on humans—though the degree of dependence may vary greatly from culture to culture around the world, something we still know little about. It is also a fact that we humans selected dogs for certain traits—we reduced shyness, promoted trainability, and fostered the dog's gaze (although we cannot say whether the last selection was done consciously or unconsciously). These dog-typical abilities have thus developed in the course of domestication and are described and explained on the following pages.

Helper or Parasite?

But what kind of relationship has developed between humans and dogs during their long time together? Dogs are dependent on humans. But is perhaps the converse also true? Are we also dependent on them? It is indisputable that even today dogs can make our lives immensely easier. We make use of their fine nose, their quickness, and their social skills.

We still use them for hunting, as watchdogs, and to herd sheep. But in more recent times, many other uses have been added, such as that of the guide dog for the disabled. After very extensive training, they are able to guide blind people on the street. Others learn to assist wheelchair users in their daily activities. They open doors, operate light switches, and fetch important objects, such as the telephone (Fig. 2.2).

In the future, dogs may be used even more in the medical field. They are already being used in psychiatry, for example. As a study in Leipzig has shown, they might help in the diagnosis of mental disorders. For this experiment, children and adolescents were observed for 25 min with a therapy dog. The young patients' mental disorder was already known. Their behaviour towards the dog was now analysed. Depending on the disorder, the patients behaved differently. The autistic children, for example, engaged with the animal frequently and briefly. The patients with anxiety disorders rarely played with the therapy dog, but for longer periods of time when they did. Thus, the way they behaved towards the dog said something about their disorder. This could help in the diagnosis of such disorders in the future.

Dogs can also be helpful in diagnosing physical diseases. They can detect bladder cancer by smelling urine samples from sick people. They can even tell epilepsy patients when a seizure is imminent and diabetes patients when diabetic shock looms. Which cues the dogs use, whether they orient themselves

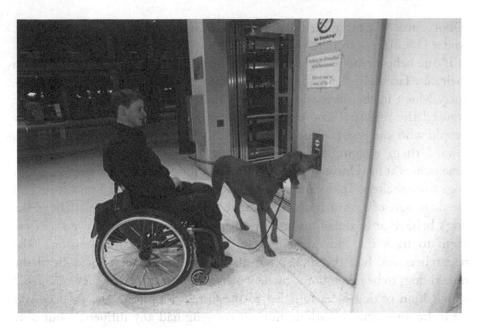

Fig. 2.2 A disability assistance dog in action

on the behaviour, posture, facial expression, or smell of the endangered person, is currently being researched. In any case, they benefit from the fact that they always keep an eye on their human counterpart, a fact that we will take up again below.

People have been making use of dogs' excellent noses for a long time, for instance, at the police or in mountain rescue services. More recently, dogs have been used not only to detect living or dead people, explosives, or drugs, but also in nature conservation and species protection. In forests, so-called beetle-sniffing dogs detect the harmful Asian long-horned beetle and thus help to quickly identify infested trees. Dogs are also trained to track lynx, hedgehogs, snakes, bats, birds, wolves, and otters, and are thus used to monitor the populations of endangered species. There are even detection dogs used on boats to sniff out whale droppings in ocean water. Since recent studies show that the use of dogs is highly effective in monitoring populations, we can expect the use of such conservation dogs to become even more widespread.

Thus, dogs still benefit us today in a number of ways. However, there are scientists who put forward a provocative thesis: They claim that man's best friend is an extraordinarily successful parasite. That is, an animal that feeds and reproduces at the expense of another living being—in this case, humans. The *cost* of a normal family dog would far exceed its *benefit*. In fact, about 4.5

billion euros is spent annually on the 6.9 million dogs that exist in Germany. And a U.S. American invests on average US$502 per year in his dog.

If one believes the parasite thesis, then our housemates manipulate us with their cute faces (Fig. 2.3). They make us care about them through their appearance. Much in the same way that their ancestors, the wolf pups, may have aroused the maternal care of mothers of the time. There are certainly few people who can resist a puppy dog with its big round eyes when it comes toward them wagging its tail. But adult dogs also arouse our affection. Researchers at the University of Portsmouth studied what makes people adopt dogs from shelters. To do this, they filmed 29 shelter dogs for 2 min each. (All the dogs were of a similar, mastiff-like breed.) The scientists then analysed the dog's behaviour in detail, using a special coding program that even allowed them to track the individual muscles that control facial expressions. The researchers also had the shelter tell them how long it took for each of the filmed dogs to be taken home by new owners. From this data, they calculated which kind of dog behaviour led to the fastest adoption. The surprise was great, because neither barking nor tail wagging had any influence, but only the 'sweet' dog gaze. More precisely, the raising of the inner eyebrows decided the dog's fate. Dogs that raised their inner eyebrows particularly often were adopted more quickly! In the interim we have learnt that dogs primarily display this facial expression when we humans are looking at them (see also Chap. 7). Even more interesting is the comparison with their wolf relatives.

Fig. 2.3 Do dogs provoke our feelings of care with their cute faces?

For wolves do not even possess this particular muscle. They cannot raise their inner eyebrows even if they wanted to. Thus, the proverbial dog's gaze is something that has evolved in the course of domestication. We humans prefer dogs that have a particularly sweet gaze.

That does in fact speak a bit in favour of the parasite thesis. We take care of dogs instead of having our own offspring! We do, in fact, play with dogs like children. And we do, in fact, pet them and talk to them in baby talk. Moreover, it is believed that in the course of domestication, humans gave special encouragement to those dogs that behaved like human children. And maybe dogs are sometimes a replacement for children. This is supported by a study in which ten families with and without children were compared. Each had a dog. The childless adults engaged with the animal much more often and intensively than the parents did. So is our dog really a parasite? Is the gaze from cute round dog eyes pure manipulation? Does the dog with this gaze induce us to invest a lot of time and money to solve its problems? Well, even the proponents of the parasite thesis have to admit that it is difficult to measure the *benefit* of such a pet. Many studies show that dog owners live healthier lives. This is not only due to the fact that they regularly go for walks in the fresh air. It has also been proven that they are less susceptible to stress and certain heart diseases. So much so that dog owners are eight times more likely to be alive 1 year after a heart attack than people without dogs! Simply spending time with an animal lowers blood pressure and heart rate. It has also been proven that older dog owners go to the doctor less often than people of the same age without a dog. For Germany, there are even concrete calculations that pet ownership resulted in the reduction of the number of doctor visits by about 2%. This reduces the expenditure for the health system by 1.5–3 billion euros, and thus provides a dog-based benefit that can even be quantified.

We also benefit from dogs on another level. They facilitate contacts with other people. This has also been shown in a series of scientific tests. In each of these tests, a test subject went walking with a dog. In the control situation, the test subject walked the same path—but without a companion. They then looked at how often this person came into contact with strangers. Whether in the park or in the city, whether the dog was well trained or not, there were always positive reactions from other people. There were no such reactions when the person was alone. In one of these studies, wheelchair users with a dog were approached eight times on their rounds, but without a dog only once. In the company of a Labrador, the test person was smiled at more often than with a Rottweiler, but this also facilitated contact with strangers. Surely every owner can confirm this. How often are you approached in the park:

'Male or female?', 'How old is it?', 'Is that a golden retriever?'. You can quickly strike up a conversation.

Our four-legged companions increase self-esteem and ease stressful situations. They bring developmental advantages to children who grow up with them (Fig. 2.4). There are even studies that indicate that bereavements in the family are easier to overcome when a dog lives in the household. Thus, the benefit for us humans from just the presence of animals can hardly be measured. Many hospitals and old people's homes have also recognized the benefit of employing dogs and now use them as four-legged therapists.

Fig. 2.4 A dog in the family promotes child development

A Dog Is Still a Dog

Even if the benefits of a normal family dog are often underestimated, the proponents of the parasite thesis are certainly right about one thing. We are prepared to invest a lot for our four-legged friend. Whether our love is always good for them is doubtful. Certainly, it does the pug no harm if it is showered with cuddles. Similarly, the old lady with a poodle, who is convinced that her pet understands 'every word' can have a nice life. A scientific survey of owners from Pennsylvania, USA, has even proven that people with a particularly close bond to their dogs also consider them to be particularly smart.

But exaggerated 'dog love' shows excesses of a completely different kind. For example, there is dog wellness or the option of having the ashes of a deceased companion pressed into a diamond. Cosmetic surgery for dogs is also offered. Particularly popular, for example, are silicone testicles, which are said to restore a castrated male dog's sense of masculinity. It is unclear whether some forms of humanised care actually do more harm than good to the dogs concerned.

However, the damage to dogs of channelling love through their stomachs has been proven. Since their domestication, dogs have been predominantly waste-eaters or self-sufficient; in many parts of the world they still are today. Obviously, they have survived very well in this way. This does not stop many in the Western world from spreading a lot of theories about the ideal, the truly appropriate dog diet. But no matter whether their mistress or master gives them dry or wet food or 'biologically appropriate raw food', most dogs today are too fat. It is estimated that 40% of pets in the industrialised countries of Central Europe are overweight. In Germany, there are even solid figures: A study by the University of Munich speaks of 52% of dogs weighing too much, while in the USA the figure is as high as 56%. And this is where the real damage begins for the dogs: excess pounds can reduce life expectancy by up to 2 years. As with humans, diabetes, arthritis, and heart disease commonly accompany overweight.

But that's not all. Humanisation creates further problems. Not only are demands made of dogs that they cannot meet, they are often misunderstood. Because even though we may sometimes feel that way, we are not born understanding dogs. We have to learn dog language first. A whole series of new studies have asked how well people understand dogs. To find out, test subjects are shown different pictures or video clips or played sounds made by dogs. The test participants are asked to assess whether the dog in the recordings is more likely fearful or aggressive or happy. All these studies show that people are not very good at understanding dogs' body signals. Apparently, not all dog emotions can be identified equally well, with humans being better at recognising positive emotions such as joy and often being wrong about negative emotions like fear. It can be dangerous when children mistake angry or aggressive dog faces for happy ones and have trouble recognising when a dog is fearful. Surprisingly, people who own dogs are not necessarily better at correctly assessing dog emotions. In some studies, they even performed worse than contemporaries who have little or no contact with the animals.

We are all well-acquainted with the phrase: 'It just wants to play'. When we hear it, we almost expect something to happen. All too often, dog owners misjudge their pets. And such misunderstandings can lead to biting incidents, which invariably cause a scare. We seem to have a certain image of 'man's best friend', of our loyal companion. We expect it to always be friendly and playful. But if, in our eyes, it is not, if it defends itself, its territory, or its food, or if it follows its hunting instinct, then a dog is acting just as appropriately for its species. For many people, however, the dog is then seen as unpredictable and aggressive.

However, biting incidents can be avoided if basic rules of behaviour are observed. A whole series of studies have evaluated extensive statistics on biting incidents in recent years. They all find similar patterns. The victims are primarily males or children between 5 and 9 years of age. The children are more likely to be injured in the head and neck area, whereas the adults are more likely to be injured in the extremities. The dogs that bite are mostly male, unneutered, and of medium size. It is extremely rare that dogs are trained to attack humans. Often, however, social insecurity and breed-inappropriate rearing play an important role. It is not uncommon for dogs to bite under special circumstances, for example, when humans intervene in fights between dogs. Interestingly, and this may sound surprising to many, dogs and their victims usually already know each other. It is rather rare for unknown dogs to attack someone who is out walking. It is also relatively rare for stray dogs without owners to bite.

One question that is often asked in this context is whether certain breeds are particularly dangerous. A large-scale questionnaire campaign conducted in the USA with more than 1500 participants brought surprising results to light. According to the survey, the dogs that are most likely to snap are rather small. Dachshunds, chihuahuas, and Jack Russell terriers bite both their masters and strangers. Australian cattle dogs are particularly aggressive towards strangers, while cocker spaniels and beagles often snap at their owners. As a rule, sheepdogs are involved in a large number of biting incidents. This is certainly also due to the fact that they are very widespread. It turns out that retrievers, on the other hand, are decidedly peaceful.

With precise data on the risks, targeted preventive measures can be suggested. For example, the socialisation of dogs is important, that is, their contact with other people and dogs in the first year of life. Because dogs must also learn how to behave socially. Animals that have attended a puppy class act more socially and have been shown to bite less often. Basic obedience training and regular exercise are equally helpful. It is important to give dogs the exercise and variation they need. It is well known that there are significant differences between breeds. A Border collie has completely different requirements than a Bernese mountain dog or a Labrador retriever. On the other hand, the repeatedly demanded mandatory leashing is certainly not a good recipe against dog bites. A study from Kiel proves how important exercise, a variety of stimuli, and contact with other dogs are. A dog that has to walk one and the same path on a short lead every day is not only restricted in its natural movement. It can also only move on a route that its owners select according to exclusively human needs. In this way, it does not learn how to behave towards other dogs. It also has no opportunities whatsoever to make contact on its own and to gain experience with its environment. And dogs are interested in this, just as we humans are.

However, it is not so much the dogs as the owners that are decisive in preventing biting incidents. The better dog owners know and understand their animals, the better they can anticipate critical situations. This is because many dogs do not react aggressively in general, but only in very specific situations. Statistics also indicate that dogs given as gifts bite more often. Perhaps dangerous dogs are simply given away more often. Perhaps the owner's understanding of his or her dog plays an important role here too. One is certainly less prepared for a dog when given it as a gift than for a dog one has acquired on one's own. Children are often victims of biting incidents because they often misunderstand dogs, especially when the latter show fear. This is not to say that children should not play with dogs. On the contrary: dogs are good for children. But especially small children should only be with dogs when supervised. In addition, the behaviour of their four-legged friends should be explained to them again and again. Because even if people do not intuitively understand dogs, they can learn to interpret the animals correctly and avoid biting incidents.

That means we have to keep in mind—our companion is not a human being, it is 'only' a dog. On the one hand, this means that it is descended from the wolf, a social predator that defends its territory. But above all, it means that it views its environment through canine eyes and behaves according to its own rules. Our task as owners is to understand these rules. It is true that the dog has developed some special skills in the long time it has spent together with humans. It is very well adapted to live in our world and to develop a very close relationship with us. However, it is naturally not capable of taking the lead in the human world.

We can only hope that our millennia-old relationship with the dog has a future. Even though circumstances have changed—and we no longer roam as hunter-gatherers—the dog is fantastically adapted to living together with us. And this coexistence is quite possible—if we also try to understand the dog and treat it according to its own nature.

3

Dogs Are Not Wolves

Ernstbrunn is in Austria, about 40 km north of Vienna. The small town is home to a very typical wildlife park. Spacious enclosures in oak woods filled with mouflons, wild boars, and Alpine ibex. There are also several wolf enclosures here. But what is that? Those aren't wolves at all, those are domestic dogs running through the enclosure. The dogs are kept here in the Wolf Science Center, founded in 2009, just like their wild cousins. Because in Ernstbrunn, scientists are looking for similarities and differences between dogs and wolves. They want to find out exactly what makes a dog a dog and what distinguishes it from a wolf.

Because normally the comparison between dogs and wolves is always a bit lame. For example, if you test wolves in a zoo and dogs in a flat, you are dealing with animals that have grown up in very different worlds. If they behave differently in experimental situations, it could simply be because each has had different experiences. Differences in behaviour would therefore not be due to differences between species. The solution to this dilemma, which was found in Ernstbrunn, is to let representatives of both species grow up in exactly the same way (Fig. 3.1).

Here there are two options. One could let wolves grow up like dogs, i.e., try to accustom them to collars and street noise. This has indeed already been tried, for example, by scientists in Budapest. This resulted in a few interesting studies, but overall, it didn't go very well. Because the wolves of today are shy and easily startled and can hardly be trained like dogs. As mentioned above, wolves were hunted to extremes. Only those survived that avoided humans at all costs. And these animals are naturally not well adapted to life in a large European city. In Ernstbrunn, therefore, the second option was chosen. They

J. Bräuer, J. Kaminski, *What Dogs Know*, https://doi.org/10.1007/978-3-030-89533-4_3

Fig. 3.1 Comparison of dog pups and wolf pups

keep dogs like wolves, in packs, in an enclosure. This is probably not quite as ideal for dogs, who prefer humans as social partners. But in Ernstbrunn, both wolves and dogs have a lot of contact with the keepers, assistants, and scientists due to an intensive programme of studies.

But what could these studies show us, where are the crucial differences between dogs and their non-domesticated cousins? When comparing the two species, perhaps the first thing that leaps out at you is the astonishing diversity of forms in dog breeds. True, wolves also differ considerably in size. The Arabian wolf, for example, weighs only about 20 kg, whereas in the Arctic, animals have been found that weigh in at 80 kg. But only in domestic dogs do we find this variability in shape, colouring, and coat, caused by breeding. Dogs can also have young twice a year. Wolves have offspring once a year at most.

There are also anatomical differences. A dog's set of teeth is smaller in relation to its body weight than that of a wolf. Its brain also weighs about 30% less. The smaller size is found above all in those areas of the brain associated with the eyes, ears, and nose. For this reason, it is assumed that most dogs cannot see, smell, or hear as well as their wild relatives. But the changes that interest us most in this chapter are those related to their behaviour and cognitive abilities.

In the Pack

Wolves, as we know, live in packs. But street dogs and the 'enclosure dogs' of Ernstbrunn also form such social groups. In one study, it was observed how wolves and dogs form in-group hierarchies and share food with each other. When prey was placed in the enclosure, the dominant dog tried to defend the food. The subordinate dogs usually accepted this, and the dominant dog was allowed to keep the meal for itself. By contrast, subdominant wolves did not let themselves be driven away by the alpha wolf. They too ate from the prey.

The fact that individuals have to compete for food is a typical disadvantage of living together in a group. This is true not only for wolves, but for all social animals. But what happens if there is a serious quarrel? Then it would be important to reconcile again, to make up, as soon as possible. Only then can the group continue to stay together and cooperate with one other. In animals, we now also speak of reconciliation when former adversaries come together again after a conflict. In Ernstbrunn, scientists have now compared how wolves and dogs behave after conflicts. More than 900 h of video material have been analysed. Reconciliation was judged to have occurred when competitors sniffed each other, rubbed against each other, or lay down together shortly after a conflict. The scientists observed the following: At first glance, dogs seemed to be more peaceful than their wild relatives. They came into conflict less often. But when a quarrel did break out, it was conducted much more aggressively than among wolves. More importantly, dogs did not reconcile, they avoided each other after a conflict. Wolves, on the other hand, seemed to be masters at making peace. They usually reconciled within a minute after the conflict! The pack dogs, on the other hand, seemed irreconcilable. Sometimes the conflicts escalated to such an extent that individual dogs had to be removed from the Ernstbunner pack, something that happened much less frequently among the wolves.

All in all, these observations fit very well into the picture: wolves are perfectly adapted to their pack system, whereas with dogs it doesn't always work out so well. Conflicts can also degenerate because breed-typical constraints make communication difficult. For example, statistics show that large poodles with cropped tails get into fights particularly often. Perhaps this is because their opponents do not understand the signals they give.

Feral dogs also form packs, but the lifestyles of such groups vary considerably depending on environmental conditions. The decisive factors are probably the food supply and the behaviour of the people in the given region. Almost always, these dogs live from refuse and do not hunt together. This makes them very dependent on humans.

In the surroundings of the Italian capital of Rome, feral dogs have been very closely observed for years. Here they also form larger groups often exceeding 20 animals. These dog groups show typical characteristics of wolf packs. They exhibit, for instance, a hierarchy based on age differences. Usually, only certain dominant animals can reproduce. In Italy, scientists have also observed that these packs raise their young together and defend their territory against other dog packs. Perhaps these dog packs would even hunt together—like wolves. However, this is not necessary here, with the Italian dog packs feeding primarily on human refuse.

In general, we still don't know too much about how feral dogs or street dogs live in different regions of the world. There is an enormous range. They range from strays that sometimes find their way back home, to dogs living on the streets that are fed regularly, to almost feral dogs that clearly avoid humans. It is still completely unclear what factors play a role here as to whether these dogs remain loners, form small, loose groups or wolf-like, structured packs. The only thing that is certain is that dogs not only vary greatly in looks, they are also adapted to survive in the most diverse of circumstances.

Puppies and People

But what distinguishes dogs most from wolves is their strong lifelong bond with humans. Wolf pups develop faster than their domesticated relatives and can only get used to humans if they have a lot of human contact in their very first weeks. Dog puppies, on the other hand, show a special interest in humans on their own.

In Budapest, researchers wanted to know exactly who or what their 3- to 5-week-old hand-raised dog and wolf pups would prefer. To do this, they conducted a very simple experiment.

A puppy was allowed to move freely for 5 min in a room containing two different objects or living beings. The respective 'adoptive mother', the puppy's caregiver or person of reference, always sat in one corner. In the other corner was the baby bottle with milk with which the puppy was fed. Or an unknown person. Or there was an unknown dog sitting there. Or a sleeping sibling was lying there. Who or what would the animal be more interested in? The scientists observed how the pup behaved, where it ran and with whom or what it spent more time. Both species, dogs and wolves, naturally spent a lot of time with their human caregiver. The tame wolf pups were also very interested in the unknown dog. But they also ran to the strange human—especially when he or she was still completely unknown. The dog pups, however, were completely fixated on the person; in the beginning they stayed more with their adoptive mother, at 5 weeks they also approached the strange person. In contrast to the wolves, they were hardly interested in their fellow dog.

Perhaps even more interesting are the other differences that scientists found. The dogs made many more sounds—they squealed and whimpered. It is assumed that young wolves generally make fewer sounds so as not to be discovered by predators. The dogs also wagged their tails. The wolf pups did not do this at all in this test, although they do generally use this signal. The scientists from Budapest suspect that wolves only wag their tails to show submission, whereas dogs simply do it when they want to make contact with another creature. Most importantly, the dog pups looked attentively into people's faces even at this early age. We will continue to explore this oft-cited dog gaze later in this chapter.

Unhappy Alone

First, however, let us ask ourselves how this bond with humans develops and what consequences this can have. Various groups of scientists have carried out a test with adult dogs that was originally developed for children. It has to do with the bond between mother and child. The test looked at how children would behave in unknown surroundings when the mother was either present or absent. The scientists expected dogs to develop a similar bond with their caregiver as children do with their mother.

At the beginning of the experiment, the owner and dog entered a room. Later, a stranger joined them. According to a predetermined plan, the people left the room and returned. This resulted in several different episodes. The dog was either alone, only with the owner, only with the stranger, or with the owner and the stranger together in this room. Then it was observed how the dog behaved under the respective test conditions. Just like children in the absence of their mother, the dogs reacted in a visibly stressed manner as soon as their owner left the room. Even the presence of a stranger did not change this.

Now the Budapest scientists compared their hand-raised wolves with dogs. The latter had grown up under various conditions. One group of dogs consisted of hand-raised puppies. The other group tested family dogs that had grown up with their mothers.

The purpose of this comparison was to help assess the behaviour of the individual animals. These dogs and wolves were all hand-reared and had grown up in the same conditions. They therefore had the same 'personal' past. If they behaved similarly, it was due to their own development. But all the dogs tested also had something in common. They belonged to the domestic dog species. That means they have the same *evolutionary past*. If they behaved similarly and differently from the wolves, then something must have changed during domestication. What would have a greater influence in this test, the *personal* or the *evolutionary* past?

Sixteen-week-old pups were confronted with the fixed episodes. The role of caregiver was taken over by the respective 'adoptive mother' for the wolves.

The dogs were tested with their owner. The results showed impressive differences between the species. The wolves behaved similarly to both humans. They made contact with them and played with both equally. They did not show any preference for their 'adoptive mother' in the process. Only when they were left alone did they become stressed. As soon as a human was in the room, they relaxed again. The dogs were completely different. They clearly preferred their owner during the individual episodes. They sought out contact and played with him or her above all else. As soon as their owner left the room, they tried to follow him or her and stood by the door most of the time until their owner came back.

What is the explanation for the behaviour of the pups? Let's start with the wolves. They had been raised by humans. Had that simply predisposed them to humans in general? Did they not discriminate between them? Other tests in Budapest have clearly shown that hand-raised wolves can distinguish between people very well. They greet their adoptive mother much more enthusiastically than other humans. They probably differ from dogs in that they are less dependent on their caregiver.

It is interesting that *all* dogs behaved in the same way. All of them clearly preferred their owner. Their *personal* past had no influence on this behaviour. It did not matter whether they had spent their first weeks with their canine mother or with a human adoptive mother. The strong bond with their human counterpart is therefore typically dog—something that most likely developed in the course of domestication.

These findings are not surprising. After all, this is what every owner experiences who leaves home without his or her dog. The owner is often accompanied to the door and greeted stormily when he or she returns. These experiments impressively illustrate that it is unpleasant for the dog as a social animal to be without us. Although, of course, they can learn to be alone in familiar surroundings.

Fortunately, this bond with humans is very flexible. Dogs from shelters that are adopted by their new owners later in life behave the same way in the above test with owners and strangers as animals that had moved in with their owners when they were puppies. It is surprising how quickly shelter dogs take new people into their hearts. Again in Budapest, shelter dogs were divided into two groups. One group was visited by a person three times for 10 min each visit. They had never seen this person before, who then played the role of the owner, the caregiver, in the experiment. For the dogs in the second group, both people were strangers. They also showed no preference for a particular person in the test. The animals that had been visited before were quite different. Although already adults, these dogs had obviously built up a bond with this person in the short time. They clearly preferred this person to the complete stranger. They sought contact with this person and greeted him or her when this person entered the room. When this person left, they followed him or her, but not the stranger. That is, they behaved toward this visitor almost as family dogs do to their owner. Three 10-min visits were enough. The great desire for human contact made these dogs bond with a new person in a very short time!

A study from Mansfield, Ohio (USA), shows how important this contact is, especially for kennel dogs. The tested animals had each lived with a fellow dog in a kennel for a longer period of time. For the experiment, they were brought into a completely new room for 4 h. Here they were either completely alone—or they were accompanied by their fellow kennel inhabitant. In a third case, the keeper was present to care for them. Naturally, the dogs were restless and stressed in the new situation. The presence of the other dog did not help. Interestingly, however, they were measurably less stressed when their carer was sitting with them! They repeatedly sought his or her proximity and then calmed down more quickly than in other situations. These results once again show very impressively who is most important for dogs: not companions of their own species, but people.

However, as a new study shows, there are great individual differences. Most dogs are more oriented towards their owner, but some dogs also develop a very close bond with other dogs living in the same household. In summary, it can be said that dogs form a very close bond with their social partners, regardless of whether these are humans or—less frequently—other dogs. Obviously, they are very flexible to bond with different species. We take advantage of this with guard dogs, which are now once again being increasingly used with herds. They are habituated as puppies to sheep or other livestock, and then live in the herd to protect it from predators.

Wild Versus Domestic Animals

Unlike wolves, dogs develop a very close bond with humans. Researchers at Ernstbrunn have found many more differences between the two species in recent years. Dogs are better at interpreting human gestures. However, hand-raised wolves can also learn to do this. In addition, dogs are more patient and wait on average up to more than a minute for a future reward. In the corresponding test, they can choose, as it were, between the sparrow in the hand (acceptable food) and the pigeon on the roof (very tasty food, which is only available later). Dogs wait more than twice as long for this than their relatives. After all, wolves are wild animals that have to survive in the wild.

Accordingly, they have certain abilities that they need for exactly this purpose. They cannot afford to wait passively for tasty food. On the other hand, they can distinguish quantities better and understand causal relationships better than their domesticated relatives. They are more persistent when a problem needs to be solved and are more likely to take risks than are dogs. In certain situations, wolves work better with fellow members of their species than dogs do (see Chap. 9).

So are wolves better at solving problems in their environment than dogs? In a study in Flint, Michigan (USA), dog and wolf pups were again compared at solving different tasks. Overall, the wolves performed much better than the dogs did. On average, they solved six of the eight tasks, whereas the dogs did not even solve two. What does this mean? Perhaps wolves are simply greedier, more gluttonous, and thus have a greater incentive to get the food. Or they are more curious and explore their environment more attentively. Or they are more interested in objects. Or perhaps wolves are better able to understand the connection between their behaviour and its consequences. All these questions have yet to be answered definitively and will keep the researchers at Ernstbrunn busy for a long time to come.

But there is another possible reason why dogs are worse at solving problems independently than wolves. They are influenced by people. This has been impressively proven by scientists from Budapest. They used the so-called A-but-not-B test (Fig. 3.2). Imagine two barriers. The experimenter now takes a toy and hides it in front of your eyes behind barrier A. Of course, you have no problem finding it. This is repeated three times. Now the experimenter suddenly hides the toy behind barrier B. You have no problem finding the toy now either. However, small children often make a mistake here. They continue to choose barrier A, even though they have seen the object behind barrier B. This has long puzzled scientists. Do children forget the current location of the toy so quickly? Are they unable to suppress looking behind barrier A because they have been successful there so often? Now the answer seems to have been found. The toddlers were influenced by the experimenter. He or she, of course, looked at them and spoke to them when he/she hid the toy. The children had the feeling that the experimenter wanted to teach them something when he/she spoke to them and hid the toy behind barrier A several times.

As soon as the experimenter stops talking to the children, they no longer make the mistake. They then concentrate completely on the problem and can easily solve it. Now comes the big surprise. Dogs make the same mistake. But

Fig. 3.2 If the dog sees the experimenter hide the ball behind barrier A several times, it selects barrier A. If it then sees the ball disappear behind barrier B, it sometimes chooses barrier A again

just like the children, they are able to go to the correct barrier, namely barrier B, in the decisive test run if the experimenter does *not* speak to them at this point. The hand-raised wolves are quite different. Regardless whether the experimenter speaks to them or not, they focus on the problem and are able to solve it. Thus, dogs and wolves do not fundamentally differ in their cognitive abilities in this case. It's just that dogs allow themselves to be too influenced by people under certain circumstances.

This is also shown by another study from Budapest. Here, two groups of dogs that differed in their attachment to their owners were tested. One group had a very close relationship with the persons they interacted with. The dogs lived with them in the home and were part of the family. The second group were working dogs, which means they were kept for a very specific purpose. For example, they were used as guard dogs. They also did not live in the home but were kept in the yard or garden. Their owners did not have a close relationship with them.

Each of the dogs from both groups was tested with its owner. A row of food containers, which had been pushed under a fence, was placed in front of the test animal. To reach the food, the dog had to pull on the handles attached to the containers. The owner sat alongside his/her dog and waited for two and a half minutes until he/she finally encouraged it to get the food. How would the dogs solve this problem? It was measured how long it took until the tested animal engaged with the containers for the first time. Would it wait until the owner requested it to do so? Did it try to solve the problem beforehand, or did it need encouragement to do so?

The results showed big differences between the groups. The working dogs tried to get the food on their own and succeeded. They behaved independently of their owners and solved the problem faster. This is certainly not due to the fact that they were better suited for this task than the family dogs. After all, both groups contained similar breeds. The working dogs had also not received any particular training for the task. They were simply more independent.

The family dogs, on the other hand, almost always waited for their owners to ask them to do something with the containers. As a result, they were able to get less food in the limited test time. They were also generally much more dependent on their owners, always staying close to them and following them wherever they went. And they looked at their owners very often. It was solely their dependence on their owners that prevented them from solving the problem on their own.

The Dog's Gaze

In many of the studies described, the scientists noticed one thing. No matter whether 5-week-old puppies or adult family dogs were involved: Again and again, the dogs looked at the person.

To study this phenomenon, dogs were again compared with hand-reared wolves from Budapest. This time they were confronted with an unsolvable problem (Fig. 3.3). Their caregiver stood behind them while a tasty piece of meat lay in front of them in a small mesh cage. The cage was locked. The animals could see and smell the food. But it was out of reach! The young wolves tried again and again to open the cage to get to the food. They bit into the bars and even tried to dig into the box from underneath. Wolves do not expect help; they try to solve their problems on their own.

The dogs, on the other hand, gave up after a few attempts. They turned around and looked the human in the eye. They used the proverbial 'dog's gaze'. Why do they do that? And what does it mean? Basically, it is a good strategy to observe people. Because everything a person does could be important for the dog.

Fig. 3.3 The Harz fox looks at its owner when it cannot solve the problem

It is interesting that the dogs looked at the humans especially in situations in which they obviously did not know a solution. The scientists assume that the dogs use this look to make contact with humans. Do they want to tell the person something? Do they really want to ask for help in this way? These questions cannot yet be answered unambiguously with the help of these experiments. At any rate, it is to be assumed that people generally respond with a great deal of understanding to such looks. You certainly know exactly what to do when your dog sits in front of you at feeding time and keeps looking back and forth between the food container and you. Or when it stands at the front door wagging its tail because its walk has been delayed. Gazing at the human instead of trying to do it on its own? That may sound a bit lazy at first. But as you can see for yourself in daily life with your own dog or one you know, this is a very effective strategy on the dog's part to achieve its aim (Fig. 3.4)!

The proverbial dog's gaze is of much greater significance for the relationship between humans and dogs than previously thought. A Japanese research team observed glances exchanged between dogs and their owners and simultaneously measured hormone concentrations in both. They analysed the concentration of the so-called 'cuddle hormone' oxytocin. This messenger substance is found in mammals and influences social interactions. It is important in childbirth and breastfeeding and governs the bond between social partners. The role of oxytocin in bonding between mother and child has been especially well studied.

Prior to the Japanese scientists' experiment, the urine of 30 dogs and their owners was analysed to measure oxytocin concentrations. This served as the baseline against which later values could be compared. During the test, owner and dog were allowed to interact freely for 30 min, but without food or toys.

They were filmed while doing so. After the experiment was finished, the test subjects provided another urine sample. After the video analysis, the human–dog pairs were divided into groups according to how long the dog looked at the person and how often the person stroked the dog. It turned out that the dog's gaze and the human's touch increased the concentration of oxytocin in the urine of both owners and dogs.

But the scientists wanted to know more about how behaviour and hormones are connected. In a second experiment, the owners were *not* allowed to pet the dog. One half of the participating dogs was given oxytocin, the other half was not. The owners did not know whether their own dog had been

Fig. 3.4 Dogs always keep an eye on people

treated with the hormone or with a placebo solution. The results showed that those dogs that had received oxytocin looked at their owners significantly more often, and the owners' oxytocin levels also increased. In the dogs, on the other hand, no increase in the hormone was recorded this time. That is logical, because they had not been stroked. Consequently, the Japanese researchers explain the connection between the dog's gaze, petting, and oxytocin as a self-reinforcing cycle. The dog's gaze promotes the release of the bonding hormone in the owners, who in turn react with more contact, which then leads to the dogs producing more oxytocin and looking at the owner even more often. A cycle of affection. Presumably, this cycle, this close bond between humans and dogs, developed in the course of domestication. The Japanese scientists also tested hand-raised wolves in the same experiment. They looked at their caregivers much less often than the dogs did. And when they did, the oxytocin concentration in humans did not increase.

Looks Like a Wolf but Isn't One!

Now, not all dogs are the same—as all owners know, there are big differences between breeds. Many studies in recent years have looked at how the approximately 400–450 dog breeds of today differ from one another. Of course, the purpose a breed was or is used for plays a major role—this determines the abilities that the dogs belonging to this breed were once selected for. In a large-scale study, an American team identified five different breed groups according to genetic characteristics: (1) the ancient breeds—such as basenjis, Afghans, and Siberian huskies, (2) mastiffs and terriers—such as German shepherds and boxers, (3) herding dogs and sighthounds—such as Border collies, (4) mountain dogs—such as St Bernards and Rottweilers, and (5) modern breeds such as cocker spaniels and poodles.

Differences between these breed groups have been shown mainly in trainability. In a scientific survey in Hungary, almost 6000 owners were asked about the characteristics of their dogs. Ninety-eight different breeds were represented. It turned out that the genetic differences also caused measurable differences in behaviour. The old breeds were more difficult to train especially in comparison to herding and hunting dogs. The dogs from the mastiff group were braver than other dogs according to their owners. Of course, we need to be a little cautious with these results. They are only based on the statements of the owners—and these are—also according to our personal experience—sometimes not entirely objective when it comes to their own dog!

Another question is: do purely external characteristics allow us to conclude whether a breed is genetically closely related to wolves or not? Some dog breeds clearly look more wolf-like than others. In a study, researchers from Southampton asked whether dogs whose outward appearance is particularly dissimilar to that of wolves also differ from their wild relatives in behaviour and body language. They checked a dog's outer appearance against the way it communicates with other members of its breed. The signals studied included growling, stiff gait, and passive and active submission. In fact, many wolf-like looking dogs used more wolf-typical signals than other dogs. For example, the Siberian Husky communicated most like a wolf. But there was also a counter-example: the German shepherd exhibited significantly fewer such signals than its wolf-like appearance would suggest.

The development of dogs was a gradual process. But a dog's appearance can only give us limited information about how far this process has progressed in a particular breed. The example of the German shepherd shows this very clearly. Once a dog breed has become far removed from the wolf, it does not help to breed it to look like a wolf again. This will not automatically bring back wolf-like behaviour.

The appearance of a dog breed is therefore hardly indicative of its relative proximity to the wolf. This can be explained in terms of the development of breeds. Different types of dogs, such as greyhounds, mastiffs, and flushing dogs, have existed for about 1000 years. But this does not mean that every greyhound is a direct descendant of the greyhound we see in many medieval paintings. For it is only in the last 100–200 years that attention has been paid to which dog mates with which. That is to say, it is only since the nineteenth century that today's dog breeds have had breed standards under the control of clubs. These standards pertain to certain behavioural traits, such as the hunting instinct, as well as certain physical characteristics, such as coat colour. Therefore, the traditional system of breed groups, as we know it, for example, from the Fédération Cynologique Internationale, does not necessarily correspond to genetic classification. Resourceful breeders have selected for certain functions and a certain appearance and have, every now and then, introduced dogs of other breeds into the mix.

Smart Dachshund, Stupid St Bernard?

It is of course legitimate to divide dogs into breeds not only due to their genetic characteristics but also according to their function. This makes sense, for example, if you want to study behavioural differences between individual breeds. We are often asked which breeds are especially smart. The Border collie is certainly among them. Others ask us which breeds are particularly stupid, the Chow-Chow perhaps? How should we answer this question? There is certainly evidence of breed differences in temperament and motivation. For example, it has been scientifically proven that herding dogs are very easy to train and terriers are particularly courageous. Herding dogs have been selected for their ability to work well together with humans. Terriers, on the other hand, tended to work independently when driving foxes or badgers out of their underground lairs.

One question you should definitely ask yourself when deciding which breed to choose is: what was this breed originally bred for? In a study in Stockholm, German shepherds and Labrador retrievers were compared. The researchers confronted our four-legged friends with a whole battery of behavioural tests (Fig. 3.5). The dogs encountered a wide variety of situations. Oversized cardboard figures appeared out of nowhere at the side of the path or a metal ladder fell over with a loud crash. It was documented whether the dogs reacted aggressively to the attack of a stranger and whether they chased after a fast-moving object. The study showed clear differences in behaviour between Labradors and German shepherds. For example, the Labradors demonstrated stable nerves and more courage in frightening situations. They also reacted more calmly to gunfire and were generally friendlier towards people than the shepherds were. The latter displayed more defensive behaviour, which is not surprising since they were selected for their protective behaviour.

Fig. 3.5 To investigate behavioural differences between breeds, dogs are subjected to a whole series of tests

Thus, we can assume that there are scientifically proven characteristics typical of a breed. But is this also true for basic cognitive abilities? Is a St Bernard dumber than a dachshund—or vice versa? According to all we know, this is not the case. An example: As will be explained in detail in Chap. 6, dogs are excellent at utilising the human pointing gesture to find food, for example. A so-called meta-analysis of many different studies in which this pointing gesture was investigated found no differences between different breeds. Only occasionally do working dogs perform better than certain dwarf breeds. Conversely, even supposedly smart dogs have their difficulties with certain tasks. For example, Rico, the supposed 'Einstein of the dog world', could not find hidden food based on a sound. Like all the other dogs tested, he obviously did not understand the connection between cause and effect, namely that food in a shaken cup produces a sound (see Chap. 8).

Nevertheless, cognition also has something to do with trainability, curiosity, self-confidence, and motivation. And, of course, there are great individual differences, independent of breed. Some dogs are very persistent when it comes to solving a problem. Others are not. They may immediately look to their owner, probably for help. This is why, for some years now, people have also been talking about personality traits in dogs. Scientists have developed personality tests for dogs. These, in turn, consist of individual test batteries that are run through by thousands of dogs. Points are awarded for the dogs' behaviour according to certain criteria. From this, scientists calculate—in much the same way as in human personality psychology—which of the behaviour patterns are most likely to belong together. For example, is a particularly friendly dog also particularly fearless? And is a dog that is quick to defend its owner also generally more aggressive? This ultimately results in typical personality traits that can be classified along the five main dimensions of personality. For example, if a dog is *very* playful and perhaps *not very* aggressive, this is recorded. Another is *very* curious and has a strong drive to hunt.

Apart from the fact that we would like to know what personality our dog has, such research results also yield helpful conclusions for practical work. In the case of German and Belgian shepherds, for example, it has been proven that the dog's personality has a direct influence on its performance as a working dog. The owner's experience also plays a role in this. But even the most experienced owner will not be able to turn a rather lethargic dog into an outstanding working dog. Probably active, playful, and less shy dogs cope better with new situations and therefore learn more easily than more timid animals. Be that as it may, it is important to note: dogs have a personality. This is scientifically proven.

If we now include wolves in these considerations, we can sum up the findings as follows: Dogs are indeed very variable in their appearance, temperament, and personality. But the differences between individual dog breeds are by far not as great as the differences between dogs and wolves. Even if you bottle-raise wolves, you won't turn them into dogs. When you consider that the common ancestor of both species lived 35,000–40,000 years ago, this is not surprising. Since then, ancestral wolves have evolved into dogs that are dependent on humans. Into a companion that for the most part is easy to live and work with.

4

What Do Dogs Understand About Others?

A scene in the park: An old sausage sandwich is lying by the side of the path. My dog Mora has spotted it immediately and bites into it. 'Leave it!' I call. Obediently, she drops the tempting food. I praise her and keep on running. But as soon as I turn away, she is back at the sandwich. Behind my back, she has scarfed it down in no time at all.

Every dog owner is familiar with such situations. One leaves the room and immediately the dog sits down on the sofa, which is actually forbidden to it. You turn away and the dog jumps up, although it was just told to 'sit' (Fig. 4.1).

In daily dealings with dogs, the question arises again and again: Do dogs know what people are thinking? Can they guess, for example, whether they are seen by people or not? Do they know how they are 'perceived' by a person? If you think about it more carefully, this would be a great achievement: it would mean that the dog not only perceives its environment itself, but also understands that its human counterpart sometimes has a *different* perception than it does. The dogs would have to understand that a human is a being with its own thoughts, its own intentions and its own knowledge. In the sausage sandwich example, the question is, does Mora really understand that I have a different view of things when I am not looking at the sandwich?

The question of whether dogs can do this is not easy to answer. Even though it seems to me that Mora waited for the moment when I turned around, there are other explanations for her behaviour. The scene does not necessarily mean that she knows what I can or cannot see. It could be a coincidence, for example, that Mora breached the ban at the very moment I turned around. Or maybe she had simply forgotten the ban after a short time. Or perhaps she has learnt in the course of her life that there is no punishment for doing

J. Bräuer, J. Kaminski, *What Dogs Know*, https://doi.org/10.1007/978-3-030-89533-4_4

Fig. 4.1 If its owner does not pay attention, the beagle goes for the biscuits

something forbidden while not seeing my eyes or my face. To examine the question of whether Mora really understands what is going on in her human counterpart (in this case, me!), what I can or cannot see, we have to rule out all these alternative explanations.

What Do Dogs See?

To find out what dogs know about how people see, we first need to clarify a few basic questions. The first is, what can dogs see in general? If they are to understand anything about their opposite number, they must at least be able to perceive him/her and his or her eyes. Everyone has heard that dogs can smell exceptionally well. But there is surprisingly little research on their eyesight. Current knowledge comes less from behavioural tests than from anatomy and physiology, and is based on knowledge of the human eye. This, in turn, serves as the basis for inferences about dogs. This indicates that our companions' vision is much blurrier than ours. Details that they can only recognise at 20 m, we can still make out at 75 m. It is often assumed that dogs are colour-blind. However, like humans, they do possess the anatomical prerequisites for colour discrimination. They can distinguish between two different colours, but they are worse at it than humans. They recognise shades of blue and yellow, but apparently cannot distinguish red from green very well. They can also see better than we can in twilight and near darkness. Above all, they are able to detect moving objects very well. Police dogs have shown in a test that they can still detect movement at 800–900 m. You may curse this ability of your dog from time to time. For example, when it spots a hare in the field hundreds of metres away and can only be restrained with difficulty.

So my dog Mora can see me—as well as the movement with which I turn around. The next question would be whether she pays attention to where I am looking. If she is interested, then she should follow my gaze. In other words, she should look where I am looking. This situation is also common in human interaction. Imagine you are sitting in a café. The entrance door is behind you. The person opposite you suddenly looks at the door. You automatically follow his or her gaze. You turn around. Because you would also like to know who is coming.

The ability to follow the gaze of others is also found in our closest relatives. You can check this with a monkey in the zoo. Stand in front of it, look at it, and wait until it looks back. Then suddenly look up at the ceiling very attentively. Most likely it will follow your gaze and also look up. Maybe the monkey has simply learnt in the course of its life that it pays to look where others look. But maybe it really can take your point of view. If there is nothing interesting to see on the ceiling, the monkeys often take a second look at the face of their opposite number. As if they want to check, where exactly is he/she looking? Or as if they were asking themselves, why is he or she staring up at the ceiling if there is nothing there? Whether monkeys really guess what their counterparts see cannot be deduced from this simple experiment. But various other experiments have shown that at least apes can do this. For example, a lower-ranking chimpanzee prefers food that the dominant ape does not see. That means it obviously has an understanding of what the other sees and what it does not. Wolves also follow the gaze of their fellow wolves and, indeed, this ability seems to be quite widespread in the animal kingdom.

But what about dogs? You can also try the test from the zoo with your dog. Sit in front of it at eye level and look up. Unlike the monkey, it will most likely continue to look at you. But it probably won't look where you are looking. This is especially true if your dog has been trained throughout its life to pay special attention to your eyes. Dogs that pay a lot of attention to a person's eyes and that watch that person are then less likely to follow his or her gaze to the ceiling. It is a different matter if you want to let the dog know where food or a ball is by looking at it. But this will be discussed in Chap. 6. Thus, dogs only make use of the human gaze in certain situations. Does this mean that they do not understand anything about what people see?

Every animal is specially adapted to its way of life. For monkeys, it is very useful to follow the gaze of a fellow monkey. In this way, they can gain additional information about their environment. For example, they can find fruit to eat that they might otherwise have overlooked. Or they become aware of dangers. Perhaps there is a conflict within the group that they should intervene in. Or a predator sneaks up on them. Interesting things happen behind them at their sides, and also above them. So it is always advantageous for a monkey to pay attention to where the others are looking.

Dogs are also constantly watching their human counterparts. However, in a dog's life, it will rarely be the case that something relevant for it happens on the ceiling! They live on the ground, and what interests them most are food, toys, and people.

Forbidden Food

The interesting experiments are always those that have a direct relation to the way of life of the animal under study. One tries to recreate a situation that is truly relevant for the animal—in this case for Mora. Through repetition, one excludes the possibility that the behaviour occurs by chance. The design of the experiment should help explain the behaviour.

What Mora went through is an everyday situation for a dog. It is forbidden to eat a liverwurst sandwich, but then the owner is inattentive. This scene provides the idea for a study in Leipzig (Fig. 4.2). At the beginning of the test run, a piece of food was placed in front of the test dog. The person now said: 'Leave it!' That is, he or she forbade the animal from eating the food. This was followed by a series of situations in which the behaviour of the human participant varied: Either he or she left the room, or turned away from the dog, or occupied him- or herself by playing Gameboy, or kept his/her eyes closed. In each of these cases, his/her attention was *not* focused on the dog. In the control situation, however, the person watched the dog. Each test run was 3 min long. If the dog ate the food, the human made no response. If the piece of food was still there after 3 min, the person picked it up and put it back in the bag. Admittedly, this was not an easy situation for a well-trained dog. However, it had the opportunity to disregard the prohibition in the right situation. Just like Mora did in the park.

When would you defy the prohibition? Of course, only if you are *not* being looked at. In fact, the dogs behaved the same way. They often ate the food when attention was not on them. And rarely when the person was looking at them.

But what should a test dog do that has a particularly strong craving for the food? Stupidly, it is precisely under the test condition where the person is

Fig. 4.2 The dog eats the forbidden food only when the dog is no longer the focus of attention

looking at it. No chance! The room is empty, the tempting food is lying in the middle of it, and the person is looking.

The Leipzig scientists were now able to observe something interesting. When the dogs decided to eat the food in this situation, they often did it in a special way. They moved slowly and approached the food in a roundabout way. It was as if they were 'sneaking around like a cat'. We can only speculate why the dogs did this. Perhaps they were torn between prohibition and appetite? Thus, they would get closer and then move away again. Perhaps they were testing whether the ban still applied? They crept slowly to the food and listened whether the person would say 'Leave it!' again. Maybe they also wanted to avoid eye contact with the person while they were doing something forbidden? They thus turned away when they neared the food so that they could no longer see the person while eating. Maybe they actually wanted to use their bodies to block sight of the food? They positioned themselves between the food and the person, so that the latter could not see them eating it. It is still unclear for which of these reasons the dogs chose their roundabout approach.

What can we conclude from these results? Can we conclude that dogs know what a person can see? Unfortunately, it is rarely the case that a single test can answer such a question. From the test described, we can only conclude that dogs obviously can distinguish whether a human is looking at them or not. And accordingly, that they behave differently. It is particularly interesting that they can even distinguish between open and closed eyes. This requires a remarkable attentiveness of the dog to the human's face.

Domestic dogs witness owners who are distracted, turned away, or asleep on a daily basis. Perhaps these dogs have simply learnt in the course of their lives that commands need only be obeyed when the person is watching them? Or, even simpler, maybe the dogs have learnt a simple rule, based on the following association: If I can see the human's open eyes, then I must obey rules. But if I can't see the person's eyes, then I can do whatever I want.

To test this assumption, scientists from Leipzig conducted a study. They wanted to know under which conditions dogs steal tasty food that they know they should not take. After the forbidden food was placed on the floor, the way the room was lit was changed. Either the person's face or the piece of food or both were in the dark. Or everything was lit up (Fig. 4.3). If the dogs were only concerned with obeying the rule quoted above ('If I see the person's eyes, then I must obey the command'), then they would only be allowed to take the piece of food if the person's eyes were not visible. Roughly following the motto: Out of sight, out of mind. So it shouldn't matter whether the food is illuminated or not. But if dogs really understand when humans can see them, then the lighting around the food should be decisive. Because only when the food is illuminated can the person see that the dog is approaching it. In fact, the dogs were significantly more likely to steal the food when the room was completely dark. This indicates that the dogs generally took the room lighting into account in their decision. The dogs' decision whether to steal forbidden food also depended on what was illuminated. If the food was illuminated, the dogs hesitated before perhaps stealing it anyway. Only the illumination of the person's face had no effect. This means that the dogs' decision does not follow a simple learned 'eyes-are-visible-or-not' rule. Instead, the result seems to show that dogs can distinguish whether someone is looking directly at them or not.

Fig. 4.3 Dogs are more likely to steal forbidden food when it is in the dark. If the food is brightly lit and the person can see the dogs well, they hesitate. It makes no difference whether the person's face is illuminated or not

Whom Should I Beg from?

Our big question is, what do dogs know about what their counterpart is seeing? To get closer to the answer, it is helpful to examine the same issue in a completely different setup. This way we can find out if the dogs' abilities are flexible. For example, how do they behave in a cooperative situation? That is, when they don't go behind a person's back, but cooperate with him or her? Are there any indications here that they know what humans are able to see?

For example, do they make sure that people pay attention to them when they beg? Scientists from Budapest, Hungary, wanted to find out. They conducted two experiments to do so.

In the first, two people sat opposite each other at a table. Each ate a liver sausage sandwich. The dog was close to the front of the table, equidistant from both people. One of them was turned towards the dog while the other looked towards the other side of the table. In such a situation, most dogs beg, even though people try to prevent it. This is what the scientists in Budapest assumed. They asked themselves which person the tested dogs would beg from.

What did they define as begging? You are probably familiar with such behaviour. Staring at the person and the sausage sandwich. Barking. Or nudging the person. Giving a paw. Or jumping up on the person. Most people understand this behaviour as begging, and sometimes give in to it.

So the people sat at the table and the test dog was let off the lead. It turns out that the dogs preferred to beg from the person who was turned towards them. Had they simply learnt that food was more likely to be expected from this person? In any case, they could not have learnt this during the experiment, because they were never rewarded for their begging.

The dogs distinguished between a person turned towards them and one turned away from them. Would they also pay attention to people's eyes in this situation? A second experiment aimed to answer this question (Fig. 4.4). This time, both people sat so that they were turned towards the dog. Both had a scarf around their heads. In one case it covered the person's eyes. And in the other, it covered his/her forehead. Accordingly, only this latter person could see the dog. And it was from him/her that the animals that were tested preferred to beg.

You may wonder why this second person also wore a scarf, even if it did not cover his/her eyes. This was done to make sure that the dogs did not simply distinguish between scarf and no scarf. In this way, the scientists could show that the dogs also paid attention to *where* the scarf was—whether it covered the eyes or not.

Once again, in this new situation, the dogs proved to be aware of what the person was paying attention to. So this ability seems to be very flexible.

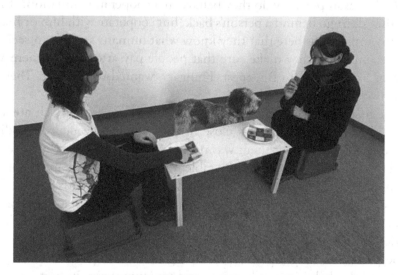

Fig. 4.4 The dog begs only from the person who can also see it

When Must I Be Obedient?

The scientists from Budapest now wanted to know whether the dogs also notice, in a further situation, whether attention is focused on them or not (Fig. 4.5). This time the animals were tested with their owners. The specific question was how well they obeyed commands when their owner was distracted.

This experiment included a pre-test. The owner looked at his or her dog and said, 'Down'. Only if the dog laid down within the next 5 s did it then get tested. The test was designed to find differences in obedience. In this case, a dog that did not do what its owner said was of no interest to the scientists. The prerequisite here was that the dog obeyed this first command.

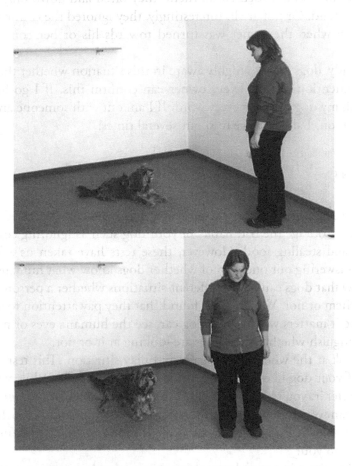

Fig. 4.5 The dog obeys the command 'Down!', but only when the owner is looking

The test situation was now as follows: The owner was talking to the testing supervisor. His or her dog was on a lead in a corner of the room. Now the owner said 'Down!'. This command was repeated two more times if the dog did not lie down.

There was now variation in where the owner looked while giving the command. He or she looked at his/her conversational partner. Or straight ahead at the wall. Or the owner looked towards the dog, in which case his/her view was blocked by a barrier. In all three test conditions, the owner's attention was *not* focused on the animal. In the control situation, on the other hand, the dog was looked at while the command was given.

When would the dogs lie down? It turned out that the dogs obeyed the command best when the owner looked directly at them.

If attention was not focused on them, they often laid down only after the third command. Or not at all. Interestingly, they ignored the command especially often when the owner was turned towards his or her conversational partner.

Obviously, dogs are also highly aware in this situation whether they are the focus of attention or not. Every owner can confirm this. If I go for a walk alone with my dog, it obeys every word. If I am out with someone and having a conversation, I always have to shout several times!

Fetch It!

It is perhaps not advisable to recreate the experiments just described at home. This may encourage the bad habits of your dog such as ignoring commands, begging, and stealing food. However, these tests have taken us a long way towards answering our question of whether dogs know what humans see. We now know that dogs can tell in different situations whether a person is turned towards them or not. We have also found that they pay attention to a person's eyes. Here it matters whether the dog can see the human's eyes or not. It can even distinguish whether those eyes are looking at it or not.

Let's look at the whole thing again in a play situation. This test is easy to recreate if your dog likes to retrieve toys. You throw a ball. While your pet is running after it, you turn your back. Will it still put the ball *in front of* you so that you can see it? To do this, it must first run around you. If it behaves like the dogs tested in Budapest, it will rarely do that. It will often put the ball down behind you.

Does that mean it doesn't know that you can't see the toy this way? Probably dogs are just not used to their owner turning their back while playing ball.

And obviously they have difficulty adjusting flexibly to this new situation. Nevertheless, some dogs will bark when they put down the ball behind their owner. Maybe they want to draw the person's attention in this way?

The scientists from Budapest developed another test in which the dog was supposed to retrieve something. The owner sat on a chair. He or she was either facing the dog or facing away from it. Now the experimenter gave the dog an object in its muzzle. It was a personal item of the owner's. The experimenter said to the dog: 'Fetch it for the owner!' And the owner repeated, 'Fetch it!' The question was again where the dog would take the object. This time the dogs almost always put the object *in front of* the owner. They also did this when the owner was sitting turned away from them and the dogs had to walk around him or her. Admittedly, it took them longer. But it indicates that they again understood something about the person's attentiveness. After all, the person has to see the object if he or she is to receive it.

Let's summarise once again what the studies described above have shown us. Dogs obviously understand when people are paying attention to them. They are more likely to eat forbidden food and less likely to obey commands when the person is distracted. They prefer to beg from someone who is facing them and put objects in front of their owners. Dogs show this sensitivity to attention in very different situations. They always behave accordingly. This ability seems to be very flexible.

The question now is whether the dogs can really put themselves in our shoes. Perhaps they simply assume for themselves: 'I see the human's eyes, so I must be obedient.' Or 'I have to beg where the eyes are.' Researchers from Leipzig wanted to find out whether dogs actually understand something about our way of looking at things (Fig. 4.6). To do this, the dogs were again asked to fetch objects for a person. The situation was as follows: The dog sat on one side of the room and the person on the other. Between them were two toys. For the dog, both toys were equally visible. The person, on the other hand, only saw one toy, because there were two small barriers between him/her and the toys: a transparent one made of plexiglass and an opaque one made of wood. Now the person said, 'Fetch it!' If the dog was able to assess what the person could see, it should fetch the toy from behind the plexiglass barrier. Because this was the *only one* the person could see, so this was the only one he or she could have meant. And this is exactly what the dogs did!

Now it is possible that the dogs do not understand anything about the person's perspective on things, but simply prefer to fetch the toy from behind the plexiglass barrier. Therefore, two control conditions were established in which there was no reason to prefer the toy behind the plexiglass barrier. Either the person was situated on the same side as the dog, allowing him or

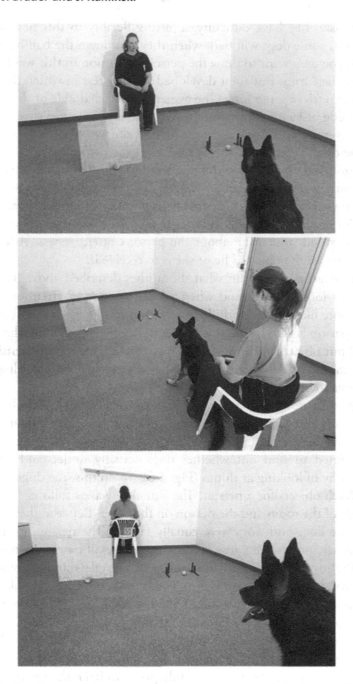

Fig. 4.6 In the first case, the person sees only one toy, namely the one behind the transparent barrier. This will be fetched for her by the German shepherd. If the person sees both toys or none (as in the latter two cases), it does not matter which one the dog fetches

her to see *both* toys equally well. Here, the dogs also showed no preference. Or the person sat in his or her original place with his or her back to the dog. Thus, the person could *not see any* of the toys. Now, the dogs did fetch the toy from behind the plexiglass barrier slightly more often than the one from behind the wooden barrier. But the preference for the toy behind the plexiglass barrier was much greater if the person could only see *one* of the two toys. This study thus confirms that dogs are really sensitive to the person's perspective on things. Out of two toys, they choose the one that the person can see. If you have a dog that likes to retrieve things, this is a test that can be easily replicated, either at home or in the park.

What Do Humans Hear?

So we can conclude that dogs are sensitive to what people can see. But there are other senses with which we humans as well as our four-legged companions perceive the environment. Hearing, for example. The question immediately suggests itself whether dogs are also sensitive to what people can hear. To test this, it again makes sense to forbid the dog from doing something. Rhesus monkeys and chimpanzees have already been studied in such situations. It turned out that these Old World monkeys and great apes are careful to keep as quiet as possible when they steal food.

A group of American scientists from Maryland wanted to know more (Fig. 4.7). They wondered how dogs would behave in similar situations. The test began with a strict experimenter forbidding the dog from eating a piece of food. Only after 5 min was the dog actually allowed to eat the food. Then two pieces of food were hidden in front of the dog in two containers. At the top of each of these receptacles, there were ribbons with small bells. In one of the containers, the clapper had been removed from the bell so that it could

Fig. 4.7 The Malinois chooses the noiseless cup and eats from it

not make any sounds. Thus, one of the containers made a sound when food was put in it or when the dog tried to get at it, but the other one did not. Now the fascinating question was, which container would the dogs eat from? After all, one would assume that the animals would try to get the food without the strict experimenter noticing.

The experiment was carried out with two groups of dogs. One group had a strict experimenter, but the latter placed his/her head on his/her knees and did not pay attention to the dogs. In this group, 18 out of 20 dogs chose the noiseless container. They obviously did not want to be heard. After all, the strict experimenter could look up on account of the noise and forbid them from eating the food again. But it could just as well be that the dogs, for whatever reason, prefer the quiet container on principle. Maybe the loud one is just scary to them. For this reason, a second group was tested. This time the experimenter observed what the dogs were doing; he or she watched them. Here the animals behaved just as chance would lead one to expect. Only half of them preferred the noiseless container. Thus, if the strict experimenter is watching anyway, there is no reason to keep as quiet as possible. This proved that dogs are not only sensitive to what people can see, but also to what they can hear.

A further study confirmed this result. In this behavioural experiment, dogs were presented with the following problem. A forbidden piece of food was placed in a small tunnel that the dogs had to put their paw through to get. This was basically quite unproblematic. Then the dogs could choose. They could approach the tunnel without making any noise by walking over a piece of carpet. An experimenter sitting opposite the dog could thus not hear the dog approaching the tunnel. However, the dogs could also walk to the tunnel over a piece of crackle foil. Of course, the experimenter would then be able to hear the dog making its way to the tunnel. The results were clear, as most of the dogs preferred the quiet side. They did this even though they could not see the experimenter while they were taking the forbidden piece of food. Obviously, they were good at remembering that a person was present. However, the dogs preferred the silent route only when it seemed really necessary, i.e., when a person was in the room and when the food was really forbidden. If the experimenter either left the room during the trial or allowed the dog to eat the food, then the dogs chose any path to the food. Thus, we can exclude the possibility that the dogs were afraid of the crackling foil. They actually understood when it was better for them not to be heard.

Seeing Leads to Knowing

Another interesting question would be whether dogs understand that *seeing* leads to *knowing*. This is not entirely straightforward. For it is one thing for dogs to understand what people see at a given moment. It is more complicated to understand that people do not quickly forget what they have already seen once. Or vice versa: that people cannot know about something they have not yet seen.

You may remember the little behavioural test in which the dogs always fetched the one of two toys that the person could also see (Fig. 4.6). Now imagine that this time each of the two toys is behind a separate wooden barrier. This means that the experimenter cannot see either of the two toys. Only the dog can see both. The experiment begins with the dog observing how two toys have been placed by a helper. What is important is what the experimenter does. He or she was present when the first of the two toys was placed behind the wooden barrier. However, the experimenter was not in the room when the second toy was placed behind the other barrier. The question now was which of the two toys the dog would fetch when the experimenter asked it to. In fact, he or she could only mean the first one, because he or she was there when it was placed behind the barrier. It is true that both toys are hidden behind barriers, but the experimenter *has seen* only the first one. The result of the study showed that the dogs did not understand what the experimenter had seen or not seen. They did not preferentially fetch the toys that he or she had seen.

French scientists from Marseille tried to answer the same question with a somewhat different experimental setup. This setup had already been used with monkeys. The question was whether dogs could distinguish between a person who 'knows' something and a person who merely 'guesses'. The experiment went like this: Dogs saw a row of four cups. Food was hidden in one; the others were empty. The dogs could not see in which cup the food had been hidden. But they could see one person (the knower) watching the hiding of the food. And they could see a second person (the guesser) standing nearby but facing away from the cups so that he or she could not see where the food was hidden. Then both people, the knower and the guesser, pointed to different cups at the same time. So if dogs were to understand that seeing leads to knowing, they should distinguish between the two people. They should follow the pointing gesture of the person who knows. For this person had in fact seen where the food had been hidden. The dogs should also ignore the cue of the person guessing. Because this person could not know anything because he or she had not seen anything. And in fact, the dogs distinguished between the two people and followed the pointing gesture of the person who knew. This showed that in this situation the dogs can certainly take into account what people have seen. It is possible that dogs understand that seeing leads to knowing. However, further studies will be necessary to really know what dogs understand here.

Our Conclusion

As always, questions remain unanswered. What owners tell us again and again has indeed been confirmed: Dogs are very good at distinguishing whether we are paying attention to them or not. They are even sensitive to what we can see from our perspective. They are not alone in this in the animal kingdom. Recent studies have shown that not only monkeys, but also many other social mammals and birds can estimate what their counterpart sees. So what is going on in my dog's head when it once again eats a forbidden sausage sandwich in the park? We now know that Mora can tell if I am paying attention to her or not. We also know that she doesn't want to be heard while approaching the food and that she's not only concerned with following the simple rule of 'out of sight, out of mind'. But does that mean that Mora gives some thought to what I'm thinking? That she really puts herself in my shoes? To understand that, science needs more time.

5

Do Dogs Learn by Observing Others?

My dog Ambula is waiting for her supper. Impatiently, she lays siege to the kitchen, with her bowl, which still has nothing in it, always clearly in view. I open the cabinet, take out the food, and fill her evening portion into the bowl. Then I put the food bag back where it came from, close the cupboard and go into the living room. Ambula eats, I read. A few minutes later I hear a loud crackling noise from the kitchen. Ambula has opened the cabinet door, pulled out the food bag and is happily munching on another portion or two. 'Smart', I think, 'she's picked that up from me'. Ambula may have looked closely when I opened the cupboard, and it may have learnt something. It certainly noticed that the food was in the cupboard. It is also possible that she picked up from me exactly how to open the cupboard and then imitated this later. But is that really the case? Can dogs learn something from others just by observing? Or was Ambula's behaviour just a coincidence?

Social Learning

As a dog owner, you often have the feeling that your dog is learning something by watching others. Be it from other dogs or from people. This process of learning by observing others is called *social learning*. We humans are, so to speak, world champions in social learning. We are constantly learning from others. In doing so, we use a form of social learning that we call imitation. Even small children around the age of 10 months begin to imitate others. In doing so, they do not simply do something that looks roughly like what someone else has done. In some cases they copy exactly what they have seen others

© The Author(s), under exclusive license to Springer Nature Switzerland AG 2021
J. Bräuer, J. Kaminski, *What Dogs Know*, https://doi.org/10.1007/978-3-030-89533-4_5

do, in other words, they imitate. Imagine, for example, that a child observes an adult operating a switch to turn on a light. Just for fun, the adult also claps his hands before the switch is pressed. So the child sees the following sequence of events: clap, operate the switch, light goes on. The clapping actually has nothing to do with the light being switched on, yet children often imitate it when they are given the opportunity to switch on the light themselves. This example should show how great the children's willingness is not only to watch what the adult is doing, but to really *copy* exactly what the adult has done. They do this even with actions that are not actually necessary to solve the problem. The children will probably even use their right or left hand depending on which hand the adult used before. So it is not only important *what* the adult did to solve the problem, but it is also important *how* the adult did it. And that is exactly the definition of imitation. Imitation means an exact copy of the 'what' and the 'how'.

Now is this perhaps a very special ability of humans? What about other animals? Isn't the expression *to ape something* used precisely because apes are so good at it? Studies with various animal species show that it is not that simple. In no animal, not even in our closest relatives, the apes, is there such a strong urge to copy the *what* and the *how* of others when they watch them solve a problem as there is in humans. So is this a very special human ability after all?

The question of whether my dog has learnt something from me by observing suddenly becomes a very important scientific question. Dogs could have developed something similar to human abilities while living together with us. That would be very practical for the dogs. They could learn a lot in a short time simply by learning by observing others. Scientists working on the question of *social learning* in dogs are primarily interested in two questions: Do dogs learn about the solution to a problem by observing others? And: Do they also copy exactly *how* someone else solves a problem?

Dogs Learn from Each Other

There are plenty of examples in everyday life that clearly show us that dogs do benefit from observing others doing something. Scientific studies also prove this. For example, a study by Austrian scientists investigated whether dogs find it easier to overcome fear of new objects when they see other dogs approaching these objects. The dogs' behaviour was compared to that of wolves. The animals were presented with new objects that they had almost certainly never seen before. In one case, the animals were alone with the new object, in another case they were with fellow members of their species. It turned out that dogs and wolves generally become more intensely involved with new objects when fellow members of their species are present. In the experiment, the animals approached the new objects more and examined them longer.

Another example are the street dogs in India's Kolkata, formerly known as Calcutta. Their puppies learn what kind of food they should prefer by watching adult dogs. This is because street dogs mostly live on human food, which is usually very rich in carbohydrates and low in proteins. Scientists observed that adult dogs in this environment strongly prefer meat when given the choice between different options. The pups in the observed groups, however, had no preference and sometimes left the meat alone. The researchers concluded that it is not innate for dogs to prefer meat. Rather, dogs develop a preference for meat after birth only over time and in social interaction.

Hungarian scientists from Budapest were able to prove that dog puppies already learn from each other at the early age of around 8 weeks. The experiment went like this: The scientists gave the dogs a box that could be opened with a small mechanism. It contained food. A first group of puppies could watch other dogs (or humans) successfully open the box to get to the food. A second group of puppies only had the opportunity to see other dogs (or humans) that were near the box without opening it. Those puppies that had observed a box being opened learnt how to operate the box more quickly than the puppies that did not have the chance to observe it being opened.

From this we see that dogs learn somewhat faster when confronted with problems in a social context. We could therefore say that dogs learn socially, for example, by imitating. But we need to be careful! Imitation is only one particular form of social learning, and one of many other possible forms. The question scientists are therefore asking themselves is how exactly dogs' social learning takes place. In a moment, we will see how scientists try to approach this question systematically.

The Fence Study

The research group in Budapest devoted a series of studies to this question. They used the fence test (Fig. 5.1). In this test, a fence in the shape of a V was set up. On the inside of the V there was food that the dogs could see. To reach it, the dogs had to run around one end of the fence. This allowed them to get inside the V and eat the food. Unlike dingoes, domestic dogs need quite a long time to reach the food if they are not helped. The scientists believe that

Fig. 5.1 Dogs find their way around the fence to the food faster if they see someone walk around the fence. But if they see a plastic car drive around a fence, that helps them just as much

one reason for the dogs' difficulties is that, in order to solve the problem, they have to move steadily away from the food. The researchers believe that the dog's desire to stay near the food may be so strong that it prevents them from solving the problem. Whatever the explanation for the dogs' behaviour in this situation, the scientists had what they wanted. A problem that was difficult for the dogs and where they would therefore benefit from help. Now it was possible to see if the dogs could *benefit* from watching someone else solve the problem.

To check this, the scientists now compared three different groups of dogs with each other. One group of dogs had never seen the solution to the problem. Another group of dogs saw a stranger walk around the fence, put down a toy or food, and walk back around the fence. Now the dogs had seen the solution to the problem. The way to the reward leads around the fence. However, to test whether the dog's behaviour was influenced by how well it knew the person involved, the scientists also tested a third group of dogs. These dogs were also shown the solution to the problem, but in this case by their own owner. It turned out that the dogs solved the problem more quickly, i.e., they found their way around the fence more quickly if they had previously seen a person go around the fence. It didn't even take them half as long to find the right way as it took the dogs who hadn't seen anything. It didn't matter if the person was a stranger or the dog's owner. Watching someone solve the problem helped the dogs.

You can try this out with your dog at home. Find a garden fence or something similar that the dog can walk around on two sides. Place a toy (or food) on the other side of the fence—initially without your dog watching you. Then let your dog watch again, and once you are sure that it has seen that there is something interesting on the other side of the fence, let it go. Now look at the clock and measure how long it takes your dog to get to the toy. After you have tried this, put the toy behind the fence again. This time, however, let your dog watch you. Thus, your dog will now see you go around the fence, put the toy down, and come back to the other side. You will now see that it is much easier for your dog to solve the problem.

In the Budapest study, the dogs' urge to take the way around the fence once they had observed a human doing so was so great that they did not even stop when a much shorter alternative was provided. After the dogs had observed the human making their way around the fence, a door was opened in the middle of the fence. This now provided direct access to the food. Nevertheless, the dogs continued to choose the way around the fence that they had learnt before. This could not be because the dogs did not understand that the door gave direct access to the food. Dogs, namely, that had not already observed a person walking around the fence chose this direct route.

Thus, once the dogs had found a way around the fence, they no longer deviated from it. This was true for dogs of the most diverse breeds and ages. After watching others, young dogs solved this problem just as well as older dogs, and dachshunds just as well as German shepherds or collies. Even the sight of another dog going around the fence helped them solve the problem. Thus, dogs also learn something by watching other dogs.

Are They Watching Closely?

Thus, dogs got a real benefit from seeing how a person or another dog solved the problem. But did the dogs also look closely at *how* the other solved the problem? Interestingly, the dogs did not copy exactly what they saw. The person sometimes walked around the fence on the right or left. Whether the dog then went around the fence on the left or on the right was completely independent of exactly what the person had previously done. So the dogs did find their way around the fence more quickly, but they did not copy exactly *how* the person had solved the problem.

At this point, scientists from Leipzig confronted our four-legged friends with a different problem (Fig. 5.2). They also wanted to investigate what a dog could learn by observing another actor. To do so, they used the following test. Food was placed under an upturned grid basket. This made the food visible, but difficult for the dog to reach. However, there was a solution to the problem. The dogs could pull a towel, on which the food was lying, out from under the basket and reach the food in this way.

What the scientists now wanted to know was whether the dogs would benefit from watching another dog—the model dog—solve the problem. Above all, they wanted to know if the dogs would copy exactly how they had watched the model dog solve the problem. To investigate this, the dog that was the model for all the others was trained to pull out the towel in two different ways. It learnt either to pull the towel out with its snout or to claw it out with its paw. One group of spectators now saw how the towel was clawed out with a paw, while the other group saw how it was pulled out with a snout. All the dogs reached the food quite quickly after the successful demonstration by the trained model dog.

But do they also copy *how* the problem is solved? The answer is no. The dogs prefer to use their paws. The dogs did this that had seen the model dog claw with its paw, as did the dogs that had seen the model dog pull with its snout. Thus, the dogs may have learnt something about *what* needed to be done to solve the problem, but the *how* didn't seem so important. This did not change even when their own owner showed them how to do it. But the scientists were now faced with a completely different question. Did the dogs really need a situation in which they could observe another solving the problem? Did they really learn by carefully observing the actions of others or did something else perhaps play a decisive role?

Fig. 5.2 Dogs are helped in solving the task just by seeing another dog eating near the basket

A Plastic Car

The research group in Leipzig subjected the dogs to another test, which yielded a surprising result. Again, the question was whether the dogs could manage to pull the food out from under the basket. And again, the dogs were able to watch another model get to the food. This time, however, there was a big difference. The model dog did not have to solve any problem at all to get the food. The food was on the towel, but outside the basket. Thus, the towel did not have to be pulled to get to the food. All the spectators saw was another dog eating food that was lying on a towel near a basket. But after watching this, the dogs were now confronted with the original problem again. And lo and behold, they pulled the towel out just as quickly as the dogs who had also observed it being pulled out. Hence, the mere fact of having seen another dog eating near the basket had an effect on the dogs' behaviour. However, this also meant that the dogs did not really care *what* the model dog did that they were watching.

If one looks more closely at the behaviour of the dogs in the fence study described above (Fig. 5.1), then there is a potentially simple explanation for what they are doing. They needn't have been really paying attention to the person or the other dog at all in order to solve the problem. Or, put another way, it may have been relatively unimportant what the other did. It was just a matter of having seen someone in the vicinity of the problem to be solved.

There is a mechanism called local reinforcement. This mechanism explains patterns of behaviour in which an animal, after seeing something or someone in a certain place, focuses more of its attention on that place and then becomes more preoccupied with it. You may be familiar with this yourself. Someone touches something and it immediately arouses your interest as well. Sometimes a place or an object becomes interesting simply because something happens in its vicinity. Imagine, for example, that you are standing in the forest and a pine cone falls from a tree. The pine cone falls onto a stone lying on the forest floor. It is only because the pine cone has fallen to the ground and thereby made contact with the stone that you notice the stone at all. Should the stone now be part of a certain problem, you have become aware of this stone through the process of local reinforcement.

Thus, simply by reinforced preoccupation with a place where the solution to a problem is located, animals may arrive at their own solution more quickly. This learning mechanism is relatively widespread in the animal kingdom. Applied to our two studies, this would mean that the dogs' attention to the basket or the end of the fence is increased—and, in this way, to the solution of the problem. This happens simply because they have seen someone nearby. In the fence study, attention given to the end of the fence is heightened simply because the dog has seen a person walking around it. Only in this way does the end of the fence become more interesting for the dogs, only in this way do they become aware of it at all. After that, it is also an easy thing for the dogs to go around the fence.

Now one could conclude: the person shows the dog the end of the fence and the dog learns something from that. Thus, the social situation exists and no matter by which mechanism, the dog learns something from it. We could now answer our question conclusively and say that dogs learn something from a social situation in this case. However, it is not that simple. Because if one starts from this simpler mechanism of local reinforcement, then it wouldn't even have to be a person who walks around the fence and alerts the dog to the solution. It would not even have to be a living creature. It would suffice, for example, to pull an object around the fence. Even then, the dog's attention would be increasingly directed towards the end of the fence.

This is exactly the kind of situation the scientists in Leipzig investigated. They wanted to know if the dogs used this simple mechanism to solve the fence problem. It turned out that the dogs solved the problem just as quickly if they had seen a small plastic car being pulled around the fence. The dogs made no distinction between a plastic car and a person (Fig. 5.1). This indicates that the dogs in the fence study may well have been following a simpler strategy. Their attention was drawn to the end of the fence by both the person and the plastic car, and then it was easy for them to solve the problem. Thus, dogs certainly do learn by observation.

This does not necessarily require a social situation involving a human. In terms of the basket study, this would mean that just the fact of seeing another dog eating near the basket serves to reinforce the dogs' attention to this part of the basket. This is how the dogs become aware of the towel in the first place and then become more preoccupied with it. By focusing more on the towel, they also find the solution to the problem more easily. Thus, the fact that another dog or a human has shown them the solution to the problem is not the decisive point. The dogs do not necessarily pay attention to the *actions* of others, i.e., not to *what* they do. The crucial point is that their attention is more focused on the basket when they notice someone or something nearby.

Let me clarify the mechanism that the dogs prefer to use here with an example. You may be familiar with this. Your dog is able to open the door. To do this it jumps up, pushes down the handle with its paws and the door is open. Maybe you are one of those dog owners who assume that your dog has copied this from another four-legged friend. Let's play out this possibility in our minds. Your dog sees another dog jump up on the door. It also observes how this dog operates the door handle and is rewarded for it: The door opens. Now it is your dog's turn. It repeats what he has seen step by step. Jump up, operate the door handle, done.

But now you may want your dog to stop opening the door. You change something. A very simple and very effective change is to simply turn the door handle around. Now you no longer have to push it down, you have to pull it up. Now imagine that there is a dog that has also mastered this method. Without any problems it opens the door by pushing the handle upwards. Your dog, who still has the other technique in its mind, sees this. What do you think your four-legged friend will do now? If your dog has really learnt something from the social situation about how opening a door works, then it should now use these new pieces of information successfully. For you as its owner, this would be no problem at all. New problem, see how others do it, new solution. For your dog, however, it is a problem. This also explains why this change in handle placement is such a useful means of keeping your dog from opening the door. Although it sees you, its owner, opening the door all the time, it is unlikely to learn this new method.

What then best explains the situation at the beginning of this example? The principle of local reinforcement. Your dog sees another dog jump up against the door. Because it is excited, your dog may jump with it. In the process, the door opens. A connection is made. Jumping up leads to the door opening. Your dog will try this out for itself and may even get its paws caught on the door handle. Then it has solved the problem itself—it learns that. But it does not learn anything about the precise connection. For this reason, it cannot transfer what it has learnt to a new situation.

One might argue that both the basket and the fence study are relatively easy problems. Maybe even *too easy*. Too easy because the dogs may figure out the solution far too quickly on their own. Maybe they don't have to learn anything from observing others. They may be a little quicker when they do so. But maybe it's still so easy that they manage the situation well without really paying attention to what others are doing.

A More Difficult Task

If you think about it carefully, it's most worthwhile to copy something by watching it being done when it's difficult or impossible to find the solution yourself. And it is especially worthwhile to see how someone else has solved a problem if you have no idea *how* to solve it yourself.

That's why scientists in Budapest subjected dogs to a somewhat more difficult problem. This problem consisted of moving a handle on a box to one side. Once this was done, a ball rolled out of the box. The dogs then had to run around the box and could take the ball, giving them the opportunity to play with it. Before the study began, the dogs were divided into different groups. For example, the dogs in one group saw the whole sequence of things: the owner pushed the handle to one side, the ball rolled out, and the owner and dog then played together with it. Careful attention was paid to the fact that some dogs saw the handle being pushed to the *left* and others saw it being pushed to the *right*. This again was supposed to show whether or not the dogs were paying attention not only to the *what*, but also to the *how*.

A second group of dogs only saw the owner pushing the handle, but no ball rolled out, so there was no playing afterwards. A third group of dogs saw only the owners touching the handle without using it. A fourth group saw the owner touch the top of the box but not the handle. To see how dogs would handle the box without any prior observation, a final group was set up. They saw the box but no touching of the box by the owner. If we recall the previous explanations of local reinforcement, it is clear that the division into these different groups was now aimed at distinguishing whether the dogs were really copying something they had observed the person doing or were again following a rather simple strategy.

If this were simply a matter of the mechanism of local reinforcement, as in the fence and basket studies, then we would expect the dogs that saw someone handling the box to be able to solve the problem more quickly than those that saw nothing at all. We would also expect that there would be little difference between the animals that saw the whole sequence of actions and those that just saw something. The dogs that saw how the handle is really used should be able to solve the problem just as quickly as those who only saw the handle being grasped and thus had it *reinforced*.

However, it is not that simple here. In this situation, the dogs seem to have really watched a bit more closely exactly what the owner did with the box. The dogs that had seen the whole sequence of actions were quickly able to solve the problem. Those who had seen the owner use the handle without the ball rolling out were just as quick. Both groups were clearly faster than all the other dogs. Even the animals that had seen the handle being touched but not operated were unable to solve the task as quickly. So it seems that here the dogs had really learnt something from watching their owners. Their awareness of the handle had not simply been reinforced because the owner had done something with the box.

Even in this last situation, however, the dogs did not look really closely at what was being done. They did not seem to care whether the owner had pushed the handle to the left or to the right. They pushed the handle to the left or to the right—irrespective of what they had observed the person doing. Nevertheless, the dogs seemed to watch more closely and learn more overall from the person's actions than in other studies. Is it possible that dogs only start to watch more closely when the task gets more difficult? When learning through observation really pays off?

Does Context Matter?

Scientists in Vienna went one step further. They asked themselves whether dogs might not only copy the behaviour of others, but even pay attention to the context in which an action is performed. They assumed that dogs might copy the actions of others especially when these actions differ from the seemingly obvious solution to the problem. This then happens according to the motto 'normally the problem can be solved differently, but he or she demonstratively did it this way, so it must also be done this way.'

The problem looked like this: A wooden box containing food was hanging from a tree (Fig. 5.3). To get to the food, a hatch had to be opened. A rope with a handle was attached to this hatch. The latter could be opened by pulling downwards on the rope's handle. If they had not observed anyone at the box before, the dogs preferred to use their mouths to pull the handle down to get the food. It became interesting with the dogs that had observed a female dog—as a model—at the box. A group of dogs observed her walking to the box with a ball in her mouth and then pulling down the rope with her paw. Pulling the rope with her paw was the obvious solution here because she had

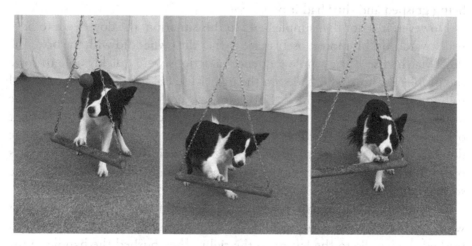

Fig. 5.3 The model dog operates the handle with its paw. Either a ball is hanging on the handle, or it carries the ball in its mouth, or no ball is visible. The observing dogs used their mouths when the ball was visible

a ball in her mouth and therefore could not use it. Another group of dogs observed the same dog walking towards the box with her mouth empty and again pulling down the rope with her paw. In this case, using the paw was not the obvious solution because the dog's mouth was free, so she could have used her mouth. Now, when confronted with the problem themselves, the two groups of dogs behaved differently. Those who saw the model with a ball in her mouth use her paw used their mouths themselves to solve the problem. Those who observed the dog without a ball in her mouth using her paw also used their paws to pull the rope. The researchers reasoned that the observers were probably using the context in their interpretation.

The reasoning goes as follows: The other dog's action is copied, but only if the action was really 'meant' in this way and there is no obvious alternative action: The scientists thus claim that dogs not only imitate other dogs, they even take into account the rationality of an action while doing so. This would indeed be remarkable, because such behaviour has so far only been demonstrated in great apes and humans.

Scientists from Leipzig, however, put forward a counter-hypothesis. They claimed that when the dogs saw another dog with a ball in its mouth, they did not then use their mouths because they considered this to be the rational activity. They did so simply because seeing a ball encouraged them to use their mouths. According to the scientists, this happens completely independently of the observed action. To prove this, the Leipzig scientists introduced a further test in addition to those described above. In this case, a group of dogs observed a model dog as it went to the box with an empty mouth and pulled down the rope with its paw to get to the food. Now a ball was present, it just wasn't in the dog's mouth. It was hanging on the box and had no influence at all on the action the dog was performing. It turned out that the observing dogs now also used their mouths, simply because a ball was visible. Whether the ball had an influence on the behaviour of the demonstrating dog did not matter to them. The behaviour of the observing dogs was quite simple to explain: Ball visible—use mouth; no other factor seemed to be important for the dogs.

From Dog to Dog

So dogs are not world champions in copying after all? Do dogs benefit only slightly from observing other dogs? What about the phenomenon with which every dog owner who has more than one dog at home is familiar? One dog has been taught a command, such as 'sit'. Now a new dog comes into the house and the owner wants to teach the new dog to 'sit' and, lo and behold, the new dog learns it much faster than the first. Doesn't it stand to reason that the young dog learnt the command so quickly because it copied it by watching the old dog?

Scientists from Leipzig based a study on this assumption (Fig. 5.4). They trained a female dog to perform certain actions that she did not have in her repertoire previously. Sometimes, for example, the dog was supposed to lie motionless on her side as if she were sleeping. This is certainly something that most dogs are not trained to do. For every action there was a command that other dogs certainly didn't know. Some commands were simply the names of colleagues involved in the trial, such as 'Joseph'. During the trial, there were so-called observer dogs. They repeatedly saw how the freshly trained dog

Fig. 5.4 A golden retriever watches the model dog lie down on its side in response to a command. Now the retriever gets the same command but cannot link it to the action

would lay down on her side on the command 'Joseph' and was rewarded for it. At a later time, the observer dogs were given the same command. The expectation was that they would follow the model dog's lead. However, this was not the case, regardless of whether the observer dog was already familiar with the action or whether it was completely new to it.

But does this mean that the new dog cannot benefit from the good training of the other dog? It may well be that a young dog benefits from the presence of a fellow dog when learning a command. This is especially the case with behaviours that dogs also do on their own, such as to 'sit'. However, this may have more to do with the general presence of another dog than with direct observation of its behaviour. If one dog is already paying attention to the person and is in a state of readiness to learn, then the other will also concentrate better on the person. It does not need to be called over, it is probably already sitting there, simply because the other dog is also sitting. This has nothing to do with imitating a certain behaviour. It is simply the social situation itself that motivates the dogs to more frequently exhibit a behaviour that they perform regularly anyway. If the dogs are then rewarded for it by the person, they have quickly learnt the link.

It is clear that for a dog's normal everyday life, a simple mechanism that does not require it to look closely is quite sufficient. Dogs don't necessarily need to learn how complicated objects like can openers or scissors work. Nor do they need to learn how to make a bed or tie their shoes. A learning mechanism that is easier than the one we humans have developed is actually quite sufficient and, as seen above, usually leads to the desired result.

We humans, on the other hand, have developed a particularly flexible form of social learning. We look closely at what others do, but we also pay close attention to how others solve a problem. It is thought that this is how we humans developed the ability to deal with such complicated objects in the first place and to develop so many complex techniques in such a short time. We look closely at how others do it and thus learn a lot in a short time that we would otherwise have had to try out ourselves in painstaking work.

The question of how and when dogs learn something from others will certainly be the subject of much further research.

'Do as I Do'

Several years ago, Hungarian scientists from Budapest published a study in which they were able to show that a companion dog for the disabled could be trained to imitate on command. The dog was shown an action, for example, the owner put a bucket on a table. He or she then looked at the dog and said, 'Do as I do'. The dog then took another bucket in its snout and placed it in the same place as the first one. This worked wonderfully in many different situations. For example, the dog did a 360° turn around itself on command if it had previously seen a person do the same. Apparently, the dog had learnt to imitate on command. The training was extremely time-consuming, and it took many hours of practice before the dog understood the general rule of the command 'do as I do'. But as soon as the rule was learnt, it very reliably imitated new actions when instructed to do so. What can we learn about dogs from this? Is this proof that dogs imitate?

To answer this question, we need to look at exactly what the dog actually learnt here. The original trigger for the apparent talent for imitation is a learned rule. The rule that the dog has actually learnt here is called the 'matching-to-sample' rule. Imagine the following situation. A dog is sitting in front of a screen. Two figures are shown to it. A yellow circle and a blue triangle. Later, it is shown a blue triangle. The dog is now asked to indicate whether this matches one of the two originals. Dogs can learn to do this. Blue triangle equals blue triangle. They may need some time for this, but they will eventually understand the basic principle and can then transfer it to new situations.

Thus, we can conclude that dogs that follow the 'do as I do' command can apply a 'match the original' rule, once learnt, to their environment and to physical actions. This is truly remarkable. But it does not mean that they imitate spontaneously.

Applications in Everyday Life

The fact that dogs are social learners can be used in many ways in everyday life. Insecure dogs can learn something from self-confident fellow dogs, young puppies can learn something from experienced older ones.

American scientists from New York have shown in a study with dwarf dachshund puppies of different ages that even very young puppies can benefit from social learning situations. Even 3-week-old puppies were able to get food more quickly if they had previously observed how other puppies had successfully completed the task. The older the puppies became, the faster they learnt. Three-week-old puppies benefited from watching others just as much as 8-week-olds. However, it was easier for the older puppies to solve the task than for the younger ones. The dogs did not lose this ability as they got older. Even though this behaviour might once again be explained by means of a relatively simple mechanism, it is impressive that it works. The pups learnt something for life just by watching.

The potential for social learning was even more evident in another experiment. South African scientists from Pretoria showed that puppies could learn to connect the smell of certain drugs with the retrieval of certain objects. Learning occurred simply by watching their mother do this repeatedly and be rewarded for it. Interestingly, the puppies that benefited most were those that had stayed with their mothers 12 weeks instead of the usual eight. But again, the dogs probably learnt this skill through relatively simple mechanisms. Puppies were introduced to behaviours only by observing their mothers, which were irreplaceable for their later training as drug-seeking dogs. This example shows yet again how effective even simple mechanisms can be. And it shows one of the many everyday uses of dogs.

In everyday life with my dog, I realise that I must be a little careful when my dog is watching me. The scene in the kitchen showed this. Maybe it would have been enough if I had accidentally run into the cupboard. If Ambula had been given food afterwards, she probably would have linked it to the cupboard door. Maybe she would then have tried out what could be done with the cupboard door and would have achieved her aim. But even an object that had touched the cupboard door would have attracted my dog's attention to it. So however much my dog really copied from observing my action, she did learn something. Maybe something for life, and now I have to find another place to keep her food.

6

How Do Dogs Interpret Human Gestures?

A scene at the lake. I throw a ball to my dog Ambula. By mistake, the ball lands not in the lake but in the bushes instead. Ambula has already swum off because, as sometimes happens, she has paid little attention to the actual trajectory of the ball. When she realises that there is no ball coming, she swims back. What now? I know the ball is in the bushes, but Ambula doesn't. But since I'm too lazy to fetch the ball for her, I give her a hint. I point to the bushes and say, 'Fetch the ball'. Ambula turns around, follows my pointed finger, finds the ball, and runs away with it. So my dog has made use of my communicative gesture, my pointing.

A scene like this happens several times a day in a dog's life. Nothing special, you will say. My dog can do that too, you will surely say. People communicate a lot through pointing gestures. And most people don't even notice. Dogs make use of these pointing gestures. This is something that every dog owner knows. It is also something that is built into dog training. For example, pointing to the toy that is to be retrieved. Or it is made clear to the dog by pointing at its basket or blanket that it should lie down in its place. Whether consciously or unconsciously, every dog is incessantly confronted with gestures from humans. So is this nothing special? Nothing that has anything to do with cognitive ability?

J. Bräuer, J. Kaminski, *What Dogs Know*, https://doi.org/10.1007/978-3-030-89533-4_6

The Object Choice Test

Various research groups around the world have investigated whether monkeys are able to use pointing gestures made by humans. To do this, they used a very simple test called the object choice test (Fig. 6.1).

Two identical cups are placed in front of the animal. Now a person hides a piece of food under one of these cups. However, the animal does not know under which of the two containers the food is. Either a barrier conceals the act of hiding. Or the person hides the food in the hand with which he or she reaches under both cups in order to secretly deposit it under one. After the

Fig. 6.1 Dogs follow pointing gestures to find hidden food

food is hidden, the person gives a clear indication as to where the food has been hidden. He or she points at the correct cup. The person does this with the intention of showing the animal where the food is. In other words, the pointing gesture is communicative. Then the animal is allowed to choose. If it chooses correctly, i.e., if it follows the gesture, it gets the food. If it chooses wrongly, it is shown that the cup is empty and that the food is in the other one.

Small children find this task child's play, so to speak. Even when they are about 1 year old, children use all kinds of communicative gestures. Which makes it all the more astonishing that monkeys, and even our very closest relatives, the great apes, fail at this task. No human communicative gesture, no matter how clear, is successfully used by the apes in the object choice test. The only exceptions are monkeys that live and have been raised in very close contact with humans. Only under certain conditions are our closest relatives able to solve a task that we take for granted and that seems to come just as easily for dogs. In fact, studies by various research groups have shown that dogs do not have the slightest difficulty using various gestures from people in the object choice test. If the person points to the correct cup, the tested dogs find the food straight away. And not only that, it is even enough to look at the correct cup. For some dogs, it is even sufficient if the human simply winks at the correct one.

You can easily recreate this test at home with your own dog. Simply take two identical cups or mugs from your kitchen and hide a piece of food or a small toy under one of them. Make sure your dog has not seen which cup the food is under. Now point to the cup under which the food is and let the dog choose. You will see that your dog will choose the correct one of the two cups without much difficulty.

Following Their Noses?

A first suspicion might be that the dogs have smelled the food under the cup and therefore always choose correctly. So they might not have followed the pointing gesture, but simply their nose. To rule this out, the scientists carried out the following test. Here, just as before, a piece of food was hidden, but the dog did not see any gesture that could have helped it find the food.

The person just sat or stood quietly between the cups without doing anything. Thus, if the dogs were simply able to smell the food, they should have also been successful in this situation and found the food. This is, however, not the case. If the dogs do not see a gesture from the person, they will not find the food. Thus, they cannot solve the problem with their nose alone.

Another fundamental question that needed to be clarified is whether the dogs really follow the cue or just the movement that the human makes. This is a big difference. We already discussed the concept of local reinforcement in Chap. 5. It was made clear how quickly an animal's attention can be directed to a place or an object simply because the animal has perceived something at that place. This can be another moving creature or a moving object. No matter what triggers the movement, it is what draws the animal's interest to a particular place.

If dogs really use pointing gestures as a communicative signal, then that would mean it is the extended pointing finger that directs them to a particular place. However, if you take a closer look at such a pointing gesture, it becomes clear that it is, of course, also a movement directed towards the cup. Therefore, there is the possibility that it is solely this movement towards the cup that steers the dog towards the food. This would then correspond to the principle of local reinforcement and would have nothing to do with the dogs using pointing as a communicative gesture. In this case, completely different movements directed towards the cup might also draw the dog's attention to it.

To rule this out, a group of researchers from Atlanta, Georgia, carried out various tests. For example, the person moved to the wrong cup, stood behind it, and from there pointed to the correct cup. He or she pointed with his/her arm, which was on the same side as the wrong cup. To further reduce the movements towards the correct cup, the arm was not fully extended, but pointed with a slightly bent arm. This was all to reduce the movements directed towards the correct cup, thus eliminating the mechanism of local reinforcement. The dogs' behaviour in this test shows that it is the pointing gesture alone that the dogs use. They continue to choose the correct cup, i.e., the one pointed to by the person, and this is the case despite the fact that the person is actively moving towards the other cup.

Thus, it is not the movement alone that plays a role when dogs follow a pointing gesture.

Exactly Which Cues Do Dogs Use?

Hence, factors such as smell and movement towards the cup alone do not help the dogs find the food. Pointing by humans seems to be an important cue for dogs. But what other forms of gestures do dogs use? And what do the dogs understand about the situation they are in? An important question scientists are looking at is whether dogs really understand pointing as 'communicative'. They do not necessarily have to do this in order to successfully utilise the pointing gesture of humans. The dogs could also follow a very simple strategy and simply choose the cup that is close to the finger that is being pointed with.

Scientists in Budapest compared different forms of pointing gestures to find out more about the dogs' strategies. The researchers discovered that the dogs had difficulty following the cues if the pointing hand did not extend beyond the human's body to a certain extent (Fig. 6.1). Thus, clear physical cues from the person, directed at the correct cup and clearly visible to the dog, were necessary for the dog to find the food. Interestingly, it seems that it must be a human hand that gives the cue. If a person stood between the two cups and pointed to the right one with a stick, as if by chance, the dogs did not follow this cue. Thus, it played a decisive role for them that it was a human hand that gave the cue. They did not just take any cue.

A study by scientists in Leipzig proved that dogs do not only search for food in the immediate vicinity of an outstretched index finger. Both young and adult dogs utilise pointing gestures to be directed to places far from the pointing hand. In the experiment, rather than being near the pointing human hand, the cups were now placed to the right and left of the dog. To succeed, the animal must therefore move away from the person's hand. Thus, if the dogs were merely to follow the strategy of searching where the person's hand is, they would not be able to solve the problem.

This also becomes clear in another Leipzig study. Here, the scientists wanted to know whether dogs distinguish communicative gestures from random movements. The random movements resembled the gesture but were not meant to be communicative. For the experiment, cups were set up again, and again the dogs did not know where the food was hidden. The scientists now compared two situations. In one, the experimenter was intentionally communicative. He or she held out his/her finger, made eye contact with the dog, and pointed to the cup where the food was hidden. While the experimenter pointed to the cup, his/her eyes moved back and forth between the cup and the dog several times. In other test runs, the experimenter made a random movement that resembled the usual pointing gesture. He or she extended his/her finger and compared the current time on a wristwatch he/she was wearing with the time on a wall clock located in the test room. Thus, the conditions in this second situation appeared similar to those of the first: there was an outstretched finger and there was a regular head movement. Only the gesture was not intentionally communicative. In fact, the dogs clearly distinguished between the two situations. They followed the gesture only when it was intentionally communicative and ignored random movements of the experimenter and his randomly extended index finger. In other words, dogs do not simply run where a finger points. Rather, they perceive gestures as communicative and know how to interpret communicative intentions.

This is also consistent with a study from Budapest. Hungarian scientists compared two situations. In one, a person looked intensely at the correct cup (Fig. 6.2). As expected, the dogs were then able to find the hidden food. In a second situation, the person also turned his/her head in the direction of the cup containing food, but then looked over the cup at the ceiling. The person's posture was similar to the first situation, but there was no communicative intention involved. The human's head was turned towards the cup as if by chance. It turned out that the dogs did not favour the correct cup in the second situation. This proved once again that dogs know how to distinguish between communicatively intended gestures and random actions.

But that is not all. Dogs also pay attention to who is making the gesture. They will follow the gesture more readily if they know from prior experience that the person making the gesture is reliable. Japanese scientists from Kyoto were able to prove this in an experiment. When dogs were subjected to the experience that an experimenter sometimes points to an empty cup, they eventually ignore the experimenter's cues. Dogs that have not been disappointed respond more reliably to a pointing gesture.

There is even more research on this topic. Italian scientists from Naples wanted to know how important gestures are for dogs. The scientists worked with a group of dogs that had learnt both a spoken command and a gesture

Fig. 6.2 Dogs can distinguish whether the human is looking at the cup or over and above it

for certain commands. When dogs in the experiment were presented with either one, the gesture or the spoken command, they obeyed. For example, they would reliably sit regardless of whether they were prompted by gesture or speech. But how would the animals react if they received both cues at the same time? And what would they do if language and gesture expressed two different commands? How would they decide if the spoken command meant 'down' and the gesture meant 'sit'? It is not possible to obey both commands at the same time. It turned out that in such cases dogs tend to follow the gesture rather than the spoken word. They sat down when they saw the gesture for 'sit', even if the owner said 'down' at the same time. The result shows once again the central importance our body language obviously has for dogs. They pay very close attention to what we tell them by means of our bodies.

From Dog to Dog

One question in this context is, of course, whether the cues need to come from people for the dogs to succeed. A group of researchers in Atlanta compared to what extent dogs distinguish whether the cue to the place where the food is hidden comes from a person or from a fellow dog. To investigate this, the scientists once again used the familiar object choice test. This time, however, instead of being in one of two cups, the food was behind one of two barriers. The test dog stood on the other side of the room and could not see what was behind these two barriers. In some of the test runs, a person was now pointing or peering behind one of the two barriers with the obvious intention of informing the dog where the food was. This is nothing new and it demonstrated once again the dogs' ability to follow these cues.

However, in some runs the scientists introduced an important change. Here, the person pointing was replaced by a dog. Due to its own interest in the food, this dog now only looked behind the barrier behind which the food was also hidden. The dog was kept on a lead to prevent it from eating the food itself. The tested dogs showed no difficulty whatsoever in deducing correctly where the food was from the interested glances of its fellow dog.

However, there are also interesting differences. Scientists in Budapest designed a test in which a film was shown on a large screen. The film showed a life-size dog. The dog to be tested sat in front of the screen. The latter received clear cues from the film dog indicating the direction in which something interesting could be seen. To do this, the film dog turned its head towards one of two plates under which food lay and stared at it. Earlier studies had shown that dogs also respond to the gesture of a human on film. But if the film dog stared at one of the two plates for a little longer, the tested dogs avoided precisely that plate and chose the other one. So if a gazing gesture came from another dog and not from a person, it was more likely to be perceived as threatening and avoided. This suggests that the ability of dogs to use human cues is a relatively recent development. It may have arisen because of the special attention dogs pay to humans.

This finding now leads us to the question: what is the origin of these abilities in dogs? Do dogs have to learn all this? Or is it perhaps at least partly innate?

Innate

In all the studies described above, there were differences between individual dogs. There were dogs that used the cues of a fellow dog better than the cues of a human. Conversely, there were animals that were better at using a human's cue than that of a fellow dog. There were also dogs that could generally make more out of a person's pointing gesture than others. If dogs behave in such different ways, it may be because they have had different experiences in their own lives. It is possible that dogs only learn in the course of their lives to pay more attention to humans and to make use of the communicative cues that the latter give.

This, of course, inevitably leads to an important question. Namely, the question whether the ability to follow human gestures has to be explicitly learnt by dogs or whether a certain disposition to follow gestures is innate to dogs. As mentioned above, every dog owner is certainly aware of how often communicative gestures are used when interacting with his or her dog. Thus, it would be quite possible that the dog simply learns to use these gestures in the course of its life, because whenever it does, something good happens: the dog finds the ball, the food, the way, etc. Even if the dog owner has not specifically trained his or her animal to use these gestures, the dog may have learnt to use them.

The question of the personal development of these abilities in dogs has been addressed by various research groups. They have investigated, for instance, the extent to which puppies already possess the ability to follow communicative cues from humans. Amazingly, puppies as young as 6 weeks use various gestures to find hidden food (Fig. 6.3). In Atlanta, scientists compared the behaviour of 16-week-old puppies that had lived with almost no human contact with that of adult family dogs and chimpanzees. It turned out that the puppies utilised the gestures of humans just as well as the adult dogs and better than the chimpanzees.

Fig. 6.3 Puppies at the age of 6 weeks can already make successful use of the pointing gestures of humans

Could it be that dogs are better at using human gestures than apes? Is it possible that they have an innate disposition to use communicative cues from people?

To answer these questions, it is interesting to note that adult dogs, as well as puppies that have grown up largely without human contact, are able to utilise human gestures. Researchers in Leipzig were able to show that stray dogs that were taken into an animal shelter from the street know how to use human gestures just as well as normal family dogs. Another study with stray dogs from the streets of Kolkata (formerly Calcutta) was able to show that the stray dogs there already use the pointing gestures of humans as small puppies.

This could be an indication that no explicit training is needed for dogs to be able to make use of human communication. This is also evidenced by a study by the Budapest scientists, in which dogs were accompanied during the first year of their lives and tested anew every month for their ability to use the communicative gestures of humans. The study began when the dogs were 2 months old and ended when they reached the age of 1 year. The dogs' ability to utilise human gestures did not change. It did not get better or worse but was consistently good. It was also shown that the type of training background a dog possessed had no influence on its behaviour. For example, dogs that took part in agility courses (in which much communication takes place by gesture) did not perform better in the study than dogs that did not take part in such courses.

But does that mean that this ability is innate in dogs? And has this ability developed specifically in dogs because dogs have lived with humans for so long? Or is it perhaps the case that canine species in general are particularly good at using communicative cues from humans?

To answer this question, it would be important to take a closer look at the behaviour of the dog's closest relative, the wolf.

Various studies show that wolves, when compared directly to dogs, are worse at using human gestures. This is true even if the wolves were raised by humans like dogs. The Budapest research group investigated this last question particularly systematically. It was already mentioned in Chap. 3 how humans had systematically raised wolf pups and dog pups. Starting at the age of 2 weeks, the pups were assigned to caretakers who took them home and with whom they lived in the same environment as a pet dog. The wolf pups were just as fully integrated into family life as the dog pups. At a certain age, both groups of pups were then tested in an object choice situation to see how good they were at using human cues. Given identical rearing conditions, the only factor that varied was the species to which the pups belonged: dog or wolf. The dogs were clearly better able to use human cues than the wolves were. This provides clear evidence that a new behaviour has developed in dogs that is not present in their closest relative, wolves.

This is, however, not to say that wolves cannot learn to follow human gestures at all. If they receive intensive training specifically designed to follow human gestures, then wolves can learn to make use of these gestures. However, they will acquire this ability much later than their domesticated relatives and will also be less flexible in dealing with human gestures than are dogs.

Three Hypotheses

One question that naturally follows is how and why the dog's ability to use communicative gestures from humans arose. Currently, there are various ideas under consideration here. On the one hand, it is possible that this ability of dogs developed through the selection of certain abilities in the common ancestor of dogs and wolves. In Chap. 2, the processes by which wolves presumably became dogs are explained in more detail. Let us now examine the various theoretical approaches once again. Wolves and humans came closer to each other. In the course of this rapprochement, it is possible that man deliberately chose the wolves that were easiest to deal with. These were certainly the animals that could be controlled and employed in a certain way, e.g., to hunt or later to herd or protect the flocks. In this way, it could be that man deliberately selected those animals for further breeding that were particularly responsive to human cues. This then resulted in a dog that had an innate ability to use gestures from humans.

This is a possible scenario, but problematic. It assumes that the people at the time had a specific plan when they came into contact with wolves. It also assumes, to some extent, that there were ways to keep wolves, even if they were not yet that tame. For people had to be able to look more closely at the behaviour of wolves to then decide in favour of one or the other. It would probably have been necessary to confine the wolves in order to be able to choose those that responded best to signs. It is highly unlikely that such possibilities existed at the time when wolves were domesticated (about 35,000 years ago). The hypothesis that humans deliberately chose wolves that responded well to human signs is therefore rather difficult to support.

However, there is also a second hypothesis. It assumes less conscious planning on the part of humans and more an indirect course of events. Let's look at the whole thing from the wolf's point of view. It is certainly the case that the wolves that approached humans had an advantage over those that did not want to do so. This advantage resulted from the fact that in the vicinity of humans they were protected from other enemies and found food that was easy to secure. One can further imagine that those wolves able to use man's gestures had an even greater advantage. They might have been able to find even more food since human gestures first made them aware of it. Or they were even better protected because they could take advantage of certain human gestures as warning signals.

Thus, it could be that over generations, precisely those wolves prevailed in the vicinity of humans that were more sensitive to human gestures than others. Thus, something that we see today in dogs would have developed rather indirectly, namely the ability to utilise communicative gestures from humans. As mentioned earlier, wolves generally pay less attention to people than dogs do. They look at humans less and are less concerned about making eye contact with humans than dogs are. This, as the studies described above on the understanding of pointing gestures also show, naturally has an influence on how wolves, compared to dogs, pay attention to communicative signals. Thus, it could be that dogs in general pay more attention to their human social partners and also more to their communicative signals.

Finally, there is a third and completely different hypothesis as to how the special abilities could have come about. It assumes that a selection against aggression and fear could have led to dogs developing these abilities. According to this theory, humans selected dogs for tameness, and then, as a by-product, the ability to utilise human gestures developed. In the human environment, those wolves that were less aggressive, less fearful, and tamer probably fared best. Particularly aggressive animals were certainly driven away or even killed, and fearful animals ran away. Only those animals that were friendly and brave were able to hold their own in the vicinity of humans. This selection alone will have led over several generations to tame animals that were free of fear and aggression. They were then able to approach humans willingly and with relatively little stress. And they certainly began to take a greater interest in humans.

Our second hypothesis indeed suggests that, to a certain extent, dogs have domesticated themselves. And exactly this is the basis of the third hypothesis. If an animal can now approach people in a stress-free way, it is only a small additional step to making increased use of the human's cues. It is through stress-free contact with humans and increased interest in them that the basis for the ability of today's dogs has been created, so to speak. In contrast to the previous theories, however, here it is not the animals that could utilise cues that were specifically and repeatedly selected for continued breeding. Rather, this ability would have arisen as a kind of by-product of something else, namely innate tameness.

To evaluate this last hypothesis, tests were carried out on foxes on a farm in Siberia. On this farm, silver foxes have been bred for several decades and divided into two different groups for further breeding. In one group, there are certain selection criteria that have to be met before the animals are allowed to continue breeding. The degree of innate tameness of these animals is assessed. Only those animals are selected for this group that are particularly free of fear and less aggressive.

For the animals in the other group, no such selection criteria hold. These animals are normally aggressive and normally fearful. A simple test was used to check the degree of tameness of the animals. For this test, a person held his hand in the animals' cage. If the animals approached without shyness and free of aggression, they were considered innately tame and were assigned to the first group. They were allowed to mate and reproduce with the other animals in this group. If the animals showed signs of shyness by withdrawing from the hand or reacted aggressively, then they were assigned to the other group.

If animals were selected according to these criteria over several generations, astonishing changes were seen in the group of tame foxes. Here, after several generations, individuals were born that showed typical dog-like characteristics, such as drooping ears, spotted and coloured fur, curly tails, and more. Selection for tameness thus had a surprising effect on the appearance of the foxes, an appearance very much reminiscent of that of dogs. And not only that. The innately tame foxes barked more than the others. They also wagged their tails more. In other words, they showed more dog-like behaviour. But do these foxes selected for tameness also have the ability to utilise human gestures? To find out, the scientists from Atlanta tested these foxes in the object choice test described above and compared them with normal foxes. It turned out that the foxes selected for tameness used cues as well as dog pups and better than the normally aggressive foxes.

But there is a problem with this study. The test for tameness—interest in the hand—on which the selection of animals is based is very similar to the test for the animals' communicative abilities. The tame foxes are, after all, the ones that approach the human hand without any problems. An interest in the human hand could, of course, have led to the successful use of the pointing gesture in this case. That is, the foxes that approach the hand are included in the tame group for further breeding. The others are not. The behaviour for which the animals were selected here is the same behaviour that they then show in the implemented object choice test. They go to the hand. The fact that this hand is pointing to a cup with food may be irrelevant.

No matter how the question of the development of these special abilities of dogs is answered in the future, the assumption that it is merely a by-product of domestication is a very interesting hypothesis. Potentially, it could be examined by studying other species that have also been domesticated. If domestication alone has an influence on these abilities, they should also be found in other domesticated animals.

Other Animals

If one compares different animal species with each other, it becomes apparent that one of two criteria must be met for animals to be able to utilise communicative gestures from humans. Either the animals are domesticated, such as goats, cats, and horses. Or the animals have received special training, as in the case of monkeys, dolphins, and sea lions. The latter all lived in zoos or animal parks and had experience with the animal shows common there, in which the animals perform various tasks in response to certain visual cues. The scientists presume that the animals learnt to apply these visual cues in other contexts, such as the object choice test.

Without such special training, only domesticated animals are actually able to utilise human communicative gestures. This applies, for example, to goats, which had grown up in a zoo with almost no human contact. Tests showed that 4-month-old young animals were already able to successfully utilise gestures such as pointing. Now one could take all this as evidence that the process of domestication alone leads to the ability to utilise human communicative gestures.

We do not yet know, however, whether other animal species can also utilise as many different gestures as the dog can. Goats, for instance, do not use the human gaze as a source of information, whereas dogs certainly can. Pigs utilise human gestures once they have gained sufficient experience with humans.

Horses utilise gestures, but not with the same degree of sensitivity as dogs do. And in the case of cats, a targeted pre-selection was required to even find animals interested in this test. Thus, the process of domestication alone is certainly not sufficient to explain how these abilities arose in dogs. It is possible that two factors played a role in the course of domestication: the selection of the friendliest animals and the selection of animals that responded well to signals. Together, they brought about what we have today: dogs that respond very well to human gestures.

Gesture as Information?

We now know which cues dogs utilise and that human gestures are of great importance to them. Dogs apparently pay attention to whether humans have a communicative intention or not. We have also seen which forms of communication are crucial for dogs, such as eye contact. Moreover, we have learnt that dogs needn't learn these forms of communication; it represents instead an innate ability of theirs that their closest relative, the wolf, does not possess. And lastly, we know that other domesticated animal species can also utilise such gestures, but that dogs are far superior to them at it.

Let us therefore move on to another issue. It can be boiled down to the following question: Do dogs understand the pointing gesture as information? What is going on in a dog's mind when a person points to a cup? Does it then think: 'Oh, there must be food there'? We now know that children as young as 1 year old understand human pointing gestures as information. Children also show things to others with the intention of informing them. When doing so, they distinguish situations in which their counterpart has seen where something has been hidden from situations in which this other person has not. If their counterpart has not seen something, children point it out to them. They inform. Would dogs do the same? Do they understand that someone who points something out wishes to inform?

Scientists from Leipzig wanted to know more about this. They conducted a study in which they compared two situations. In one situation, food was hidden and the dog could not see in which of two boxes. The experimenter then pointed to the correct box and looked at the dog sitting opposite him or her while doing so. A second person was sitting next to the experimenter. In the second situation, the experimenter acted in exactly the same way, except that he or she did not look at the dog when pointing, but at the second person sitting next to him or her. The dogs distinguished between the two situations. When the experimenter looked at the test dog, the dog followed his/her gesture and chose the correct cup. But if the experimenter looked at the second person sitting next to him or her, then the dog ignored his or her gesture and made a random choice. Now, the fact is that dogs would not distinguish between the two situations if they classified the experimenter's gesture as information. After all, the information given by the experimenter is always the same: 'The food is in the cup I am pointing to.'

This does not change, no matter who I am looking at. Children, for example, do not distinguish between the two situations. They don't care who a piece of information is directed at. The main thing is that it is information. They follow the pointing gesture no matter what. But why don't dogs do this? One reason could be that they understand the pointing gesture as an instruction, as a command. And if such an instruction is not directly addressed to them, then they can also ignore it. This would explain why dogs ignore pointing gestures when they are not directly addressed to them. The whole thing could also be understood as a perfect adaptation to living together with humans. For historically, dogs were used as social tools from a certain point in time onwards. They were willing helpers who could perform certain tasks at a distance, such as keeping a herd together or hunting a wild animal. For both activities, dogs had to be given different instructions, sometimes over longer distances. For this, it would be sufficient if dogs responded to pointing gestures as they would to instructions. They would not have to understand that gestures can have an informative character.

No matter what dogs understand about the gestural communication of humans, it remains astonishing how sensitively they respond to it. This is what fundamentally distinguishes them from wolves. So anyone who thinks that dogs are simply stupid wolves that have lost many impressive wild animal skills is mistaken. Over the course of time with humans, dogs have gained a variety of new areas of competence that wolves do not have at their disposal.

Benefits in Everyday Life

Now we can ask ourselves what benefit these skills have in our daily interaction with dogs. And surely every dog owner can think of numerous examples. For we have seen that the fact that dogs have this ability is certainly one of the reasons why they have adapted so well to life with humans. It enables the shepherd to signal his dogs over long distances to herd the sheep. The hunter is able to direct his dog without scaring off the game. Paralysed people can communicate with their dogs via eye movements, and the dogs' abilities are particularly impressive when dealing with deaf-mute people, for whom all communication with the dogs takes place via visual signals.

There is, however, at least one example that shows how much the dog profits from having this ability. For example, deaf dogs can lead a largely normal life and understand well the commands given to them if they are in the form of visual signals.

The benefits in daily interaction with the dog are great, and every dog owner communicates with his or her dog in this way. Be it consciously or unconsciously. I at least profit a lot from the fact that my dog understands my gestures, because now I don't have to crawl into the bushes looking for the lost ball. I simply point to the bushes for my dog and then it can do it itself.

7

Communication Between Dogs and Humans

A scene in the dog run. Luna, a female terrier, is barking. She barks at everything that comes towards her. People, other dogs, even birds are barked at. Soon it becomes too much for Luna's owner. 'Luna, please be quiet. Come here right now, please', he calls. Luna does not listen to him and continues to bark instead. Luna's owner intensifies his shouting. He calls louder now and with more authority in his voice. Still, Luna does not think to listen. She has quite different worries. A group of crows urgently needs to be chased away. Luna contorts her face at this. 'Now she's getting angry', Luna's owner thinks to himself. 'Luna, enough is enough and now please come over here right away', he yells. Luna doesn't give it a second thought. When Luna is just about to turn her attention to a cyclist, Luna's owner makes one final effort. There is no time to waste. 'Luna, get over here'. And lo and behold—Luna stops, turns around, and follows the command. Practically to the letter.

Dogs communicate with us humans every day like no other non-human creatures. But what actually gets through? How flexible is the communication of dogs with us and with their own species? Do dogs understand our language? What do dogs actually mean when they bark or growl? And what about facial expressions? Do dogs understand what it means when we 'make a face'? Can dogs perhaps even read our emotions? And what about the other way round? How do we understand a dog's facial expressions?

J. Bräuer, J. Kaminski, *What Dogs Know*, https://doi.org/10.1007/978-3-030-89533-4_7

How Dogs Bark

Let us first turn to the area of the production of sounds and then take a closer look at dogs. One form of communication by dogs, as the example described above has also shown, is very conspicuous. Dogs bark (Fig. 7.1). Every dog owner knows that they do this for a specific purpose. Some dog owners may also assume that they want to say something specific, such as 'play with me' or 'leave me alone'. Other dog owners may even have the feeling that their dog is saying 'Good morning' or 'Feed me'. That dogs want to 'tell' us something with their barking is true, even if only to a certain extent.

When dogs bark, they do so in many different ways. Small dogs have a higher-pitched bark than their larger brethren, simply because the vocal cords of the latter are longer and vibrate differently, similar to the low notes of a string instrument. So there are obvious reasons for different barks. But there are also other, more complicated ones. American scientists from Davis, CA, for example, have found that dog barking has different sound qualities depending on the reason for the barking. Thus, not only can individual dogs

Fig. 7.1 Dogs bark differently depending on the situation

be measurably distinguished by their barking, but one and the same dog barks in different ways. The barking of an agitated dog sounds different from that of one playing or an individual that has been separated from its pack. This is true regardless of the sex, age, or size of the animal.

Thus, dogs vary their sounds depending on the situation they find themselves in. Hungarian scientists from Budapest recorded dog growls in different contexts: when playing, when guarding, and in response to a threatening stranger. They first found that play growling was acoustically different from the other two types of growling. Then they played the sounds to other dogs (Fig. 7.2). And indeed: the growl to protect a bone kept other dogs from taking an apparently unattended bone. The other two growls did not have the same effect.

Dogs that give a play growl appear larger to their fellow dogs than they actually are. This was also the result of the Budapest study. For this purpose, two different growls were played to dogs. The first growl was recorded in a playful context, the second in a context where food was being guarded. At the same time, the dogs were each shown two photos, both of which showed the dog that had growled. In one photo, the size of the dog was correct, but in the other photo, the dog was made to appear larger. The dogs looked at the accurate picture when they heard growling being used to guard food. It can be concluded from this that the growling in this case is an honest signal. Such a signal conveys information that tells a counterpart what it is dealing with. However, the dogs looked at the picture with the unnaturally enlarged dog

Fig. 7.2 Dogs distinguish between play growling and aggressive growling. They'll approach a bone when they hear play growling from a tape, but not when they hear aggressive growling

when they heard the play growl. It may be that playing dogs exaggerate the signals they send to reinforce playful interaction.

British scientists from Sussex wanted to know if dogs could tell from the barks of other dogs whether a small or a large dog was approaching them. They played different barking sounds from different sized dogs to their participants. At the same time, they showed them two dog models. While one model matched the barking dog in size, the other was smaller. It turned out that the dogs looked at the smaller model significantly more often. Longer gaze time is a sign of surprise, and from the long gaze it can be concluded that dogs are able to correlate the barking with the photo. They can obviously draw information about the body size of other dogs from their barking.

Humans also differently perceive the different sounds of dogs. This was shown by a Hungarian study. The scientists played a tape of dogs barking to test subjects. Then they asked the test participants to judge the barking in several ways. The participants could give points for aggressive, fearful, desperate, playful, or happy barking on a predefined score sheet. No specific criteria were given to guide the participants. They were asked to judge according to their feelings. It turned out that people are not only able to distinguish different types of barking from dogs. They can even correctly judge whether we are dealing with an angry, frightened, or happy dog.

Humans can also perceive very subtle differences. In this study, test participants were presented with the three types of growls described above. They then usually assigned them to the correct emotions. Female participants were even better at assigning emotions than male participants. Participants who had experience with dogs also performed better than the rest.

A Dog Language?

Thus, dogs emit different sounds, and their barking depends on the situation they find themselves in at the moment. And we humans understand dogs to a certain extent. This leads to an interesting scientific question: Can we say that dogs have a language that is comparable to that of humans? We know that dogs have developed a range of vocalisations. These may simply be expressions of different emotional states, similar to a lion cub left alone, crying miserably, or an angry lion roaring menacingly. These would be vocalisations that have evolved for the purpose of conveying a specific message that can be understood by the opposite party. In science, this is called a signal. We can therefore state that dogs send signals that can be understood by humans and to which humans can respond accordingly.

The interesting question, however, is how flexible such signals are. To answer it, let's imagine a scene. A burglar enters a house and the dog starts barking menacingly. How can this be translated? Is the dog trying to say, 'Angry, angry, angry!' Or does it want to communicate something complex, such as, 'I want you to leave, because I know you don't belong here.' To decide this, let us return to the study in which it was shown that dogs can distinguish different kinds of growls from other dogs in different conflict situations (Fig. 7.2). Playful growling is fundamentally different from more aggressive growling. Considering this, one would have to assume that the explanation found above is correct and that different barking merely expresses different inner emotional states.

However, it is striking that the dogs in the study, which were presented with different growls from two confrontational situations, actually recognised the difference and showed clearly different behaviour in each case. This is indeed remarkable, as one can assume that the inner emotional state of a growling dog could be quite similar in both situations ('Angry, angry, angry!'). This gives us an indication that dog sounds could be more than just a signal. However, we do not yet have proof. It is quite possible that dogs have only learnt to distinguish signals and to assign them to different situations. This does not mean that they have a language.

After all, our human language lets us to interact about complicated issues without our inner emotional states playing a role. We can talk about heart-break as well as mathematics. We are able to converse about being angry and scared in the past when burglars broke into the house. We can do this even if the specific situation happened long ago and the actual feelings have long since faded. Similarly, we humans can suppress our feelings when we talk about potentially upsetting things. The important thing to remember is that human language functions independently of inner states. This is the feature of human speech that distinguishes it from dog barking.

But perhaps there has been a special adaptation between humans and dogs in this area. To find out, Hungarian scientists from Budapest used a technology new to dog research, magnetic resonance imaging. With this technology, it is possible to study the brain activity of dogs (and much more) without injuring the animals through invasive procedures. For the study, dogs were trained to lie completely still for a period of time. During the experiment, different sounds were played to them in the magnetic resonance scanner. In this way, the scientists first wanted to find out whether there are certain places in the dog's brain that react specifically to human voices. To do this, the dogs were played different sounds, vocal and non-vocal. The dogs' brain activity clearly showed that there are some areas that specifically respond to human voices. However, these areas are not activated exclusively by human language, as in the human brain, but by human sounds. Unlike humans, dogs obviously do not have brain regions that respond specifically to human language. Therefore, it is currently assumed that the dog brain has not adapted directly to human language.

Why Do Dogs Bark at All?

It is quite remarkable the variety of vocalisations available to dogs. For this reason, there are scientists who argue that the wide range of vocalisations available to dogs has developed because they have lived together with humans for so long. Accordingly, vocalisations would be an adaptation on the part of dogs to living together with humans. Is it possible that some sounds even evolved specifically for communicating with people? We have already seen on several occasions that it was just such adaptations that made dogs such excellent companions for humans. Maybe the different types of dog barking are part of this phenomenon? What we have here is the classic chicken-and-egg problem.

Which came first? We can confidently assume that the human ability to distinguish between different types of barking did not first develop as a result of living together with dogs. If you want to test yourself here, you could try to distinguish an aggressive lion roar from a friendly lion growl. This will usually go well. Likewise, everyone will be able to instinctively recognise whether a house cat is desperately meowing from hunger or a cow is mooing happily because it is standing in a green meadow. In short, there seems to be an innate human ability to distinguish and interpret certain animal calls. It is easy to imagine that in the course of evolution it was very advantageous to distinguish an aggressive animal from a friendly one. Because fleeing always costs valuable energy. Climbing a tree every time an animal makes a noise would be a very costly strategy. It would consume a lot of energy, which would then no longer be available for other things, such as raising offspring. Thus, whoever is able to gauge whether the animal in front of them is aggressive or peaceful can conserve energy. It seems obvious that humans have made use of this evolutionary advantage.

Humans therefore already had the basic prerequisites for understanding dogs very early on. There are two possibilities for further development. First, it could be that the dog's original ancestor, the 'primeval wolf', already had a variety of vocalisations at its disposal that have remained largely unchanged. With this repertoire in its pocket, the wolf gradually adapted to living together with humans. The repertoire would then hardly have changed in the course of subsequent development. We have already seen in Chap. 2, which dealt with how wolves came to be dogs, that this is not a possibility. Secondly, it could be that wolves already had a whole range of possible vocalisations at their disposal, but that the future dogs developed further new sounds in the course of their coexistence with humans. This second assumption is confirmed by studies out of Kiel, in which the vocalisations of dogs were compared with those of wolves. The scientists found that dogs use different and, more importantly, far more vocalisations than do wolves. The greater range may well represent a direct achievement of adaptation.

It could be argued, for example, that dogs have developed so many different vocalisations because humans are able to distinguish between them. If people were only able to distinguish two sounds, dogs would certainly only have developed those two. However, humans understand, metaphorically speaking, an entire alphabet. In the course of evolution, dogs that had a greater range of sounds at their disposal may even have had an advantage because they were better understood. This advantage may perhaps have consisted in the fact that they had more means at their disposal to express different needs. As a result, more of these needs could also be met by humans. The corresponding dogs were thus better at surviving, they produced more offspring, and soon there were ever more dogs that could make just as many sounds. The result of this development would be the dog as we know it today. An animal with a large repertoire of different kinds of barks.

In summary, we can say that dogs emit many different sound signals that can be understood by humans. However, these vocalisations are not comparable to human language because, according to the present state of scientific knowledge, their generation is not completely detached from inner emotional states.

Communication Through Facial Expression

Another form of communication between humans and dogs takes place through facial expressions. This refers to facial movements, which are additional signals in social situations and convey information. When we look at dogs' faces, the first thing that strikes us is how varied they are. This is, of course, primarily due to anatomical differences between breeds. Some dogs have long snouts, others short ones. Some have large eyes, others smaller ones. Some dogs have long fur, others short. But what all dogs have in common is that they move their faces and have facial expressions. The scientific study of canine facial expressions, however, faces a problem. For many years, dogs' facial movements were observed subjectively. Researchers turned their attention to the dog's facial movements and interpreted what they saw through humanised glasses. This need not come as a surprise. When people see other people's facial expressions, they can't help but understand emotional expressions in human terms. When we look at a person who pulls up the corners of his or her mouth and smiles, we understand him or her to be happy or at least cheerful. If we see a person with furrowed eyebrows, then we assume that he or she is thinking. There is no problem with this at all. Because in most cases our assumptions will prove correct. But does that help us understand dogs?

As early as the 1970s, the psychologist Paul Ekman and his colleague Wallace Friesen described a total of five basic emotions and the facial expressions that accompany them. Ekman considered these to be universal. Universal means that every person in the world, regardless of their cultural background, recognises these expressions and interprets them in roughly the same way. These universal facial expressions are joy, fear, sadness, disgust, and anger. However, we run into a problem if we transfer this human system in an unguided manner to other species. In cases of doubt, our interpretations may be completely wrong. British scientists from Lincoln asked themselves why children are so often bitten in the face by dogs. It turned out that small children often want to hug dogs, even if the dogs' body language indicates that this could be dangerous. The scientists tested a hunch. They showed photos of dogs that showed a variety of facial expressions to 4- to 6-year-old children. They then asked the children which dog they would like to hug. It turned out that the children were especially keen to hug the dog that had an aggressive snarling expression. This dog had its lips pulled particularly far back. Asked why they wanted to hug this of all dogs, the children replied, 'Because it's smiling so happily.' This was, of course, a dangerous misinterpretation. This example shows in an impressive way how difficult it can be to interpret dogs' facial expressions if we proceed subjectively.

And there is another aspect to consider: our unconscious preferences. Humans have a clear preference of which they are hardly consciously aware. They love facial expressions that are as childlike as possible, they love the so-called baby schema. In an experiment, American scientists from New York showed their test subjects photos of dogs' faces. Each subject viewed two photos next to each other. Both photos showed the same dog's face. One of the photos was slightly manipulated, a certain aspect of the dog's face having been altered. The subjects were asked to choose which of the two photos they preferred. They were given very little time to do this because they were supposed to make an unconscious decision. In the end, the subjects' preferences were clear. They preferred, for instance, the photo in which the dog's eyes were slightly farther apart. A larger distance between the eyes makes a face appear rounder, which is a classic feature of babies' faces. The test persons also chose photos in which the dog's eyes were slightly enlarged. Large, wide-open eyes are another characteristic of babies' faces. In short, the subjects made their choices with a clear preference for dog faces corresponding to the baby schema.

You may remember the study by British scientists from Portsmouth (see Chap. 2). They had shown that dogs that move their eyebrows frequently have a clear advantage (Fig. 7.3). For this study, the researchers analysed the facial movements of dogs. To do this, they used a scientific observational tool called the Dog Facial Action Coding System or Dog FACS. This tool assigns to each facial muscle a movement on the face's surface. Facial movements are assigned numbers and analysed separately. In this way a subjective interpretation of facial expressions is avoided, and an objective observation is made possible. The method is based on the Human FACS developed by Paul Ekman and follows the same basic principles.

The question scientists asked themselves after this first finding was whether dogs had perhaps learnt to move their eyebrows to manipulate humans; whether dogs consciously use their eyebrows to get more food, for example. To find out, the scientists conducted another study. In this one, a human faced a dog on a lead. In one case, the person was facing the dog, and in the other, the person had his/her back to the dog. In one case, the person held food in his/her hand, and in the other, he or she did not. Now the scientists observed the dog's facial movements. They saw that the dogs showed more facial movements when the human was facing them. The eyebrows in

Fig. 7.3 Dogs have an eye muscle that wolves do not that allows them to raise and narrow their eyebrows in a way that is irresistible to humans

particular moved. Thus, the dogs used their facial muscles in particular when a person could see them. However, it did not matter whether the person was holding food in his/her hand or not. It was not the case that food led to an increase in the dogs' facial movements. If the dogs had learnt to use the movement of their eyebrows in a deliberate way, then one would expect them to increase this movement as soon as food was involved. This, however, was not the case.

Therefore, we do not assume that dogs make conscious use of their facial expressions. But facial expressions do give dogs a selective advantage. In the course of evolution, humans have unconsciously favoured those dogs that more frequently exhibited this eyebrow movement. To see if the anatomy of dogs' facial muscles has changed specifically for facial communication with humans, British scientists from Portsmouth compared the facial anatomy of dogs with that of wolves. Based on dissections of dog and wolf heads, it became clear that the facial musculature of dogs and wolves is identical—with the exception of one muscle: the levator anguli oculi medialis. In dogs, this muscle is responsible for the intense raising of the inner eyebrow. Wolves do not have this muscle. It ensures that the eyebrow movements of dogs can be produced with the highest intensity. Thus, it can be stated unequivocally that the dog's gaze is a result of selection in dogs according to the unconscious preferences of humans.

Now, conversely, we could ask ourselves how dogs perceive the facial expression of humans and what they understand about it. Scientists from Kyoto showed that dogs certainly do recognise their owner's face. To prove this, dogs were shown photos of the faces of their owners as well as those of strangers. At the same time, the owner's voice or the voice of a stranger was played to them from a tape recorder. The dogs would sometimes see their owners' faces and hear their voices, or they would see their owner but hear the voice of a stranger, and so on. When the voice and face matched, the dogs spent less time looking at the photos. The scientists concluded that the dogs were well aware that voice and face did not match, causing them to look longer at the photo to glean more information.

A study from Vienna also showed that dogs can distinguish the faces of their owners from those of strangers. Here, too, the scientists showed the dogs photos of their owners in comparison to those of strangers. The dogs even recognised the faces of their owners when these were reduced to the inner silhouette, i.e., when hair and ears were hidden by a cloth.

Dogs also perceive the facial movements of humans. They distinguish smiling faces from neutral ones, happy ones from disgusted ones. In a study by British scientists from Lincoln, dog faces or human faces were shown. The dogs were each presented with two photos, each showing a different expression: either happy/playful or angry/aggressive. At the same time, the animals were played sounds from the same person or dog with positive or negative expressions, or a neutral noise. It turned out that the dogs spent more time looking at the face whose expression matched the sounds they were hearing at the same time. It did not matter whether the photos showed their fellow dogs or humans. These results show that dogs gain important information about the feelings of a counterpart from both sounds and facial expression. In the process, they distinguish between positive and negative expressions.

Thus, dogs pay attention to, and distinguish between, people's facial expressions. But does this also mean that dogs can recognise and read people's emotions? Recognising emotions would mean that dogs interpret or take into account a person's internal states. But do dogs really do that? The alternative is that dogs learn to associate certain facial movements with a certain context. For example, when dogs see a smiling face, they might simply have learnt to associate it with something good. Very much along the lines of 'That's the face my owner makes when something good happens.' It will be the work of scientists to try to distinguish between these two explanations. We may already have a small hint as to which might be the right one. In almost all studies, it is easier for dogs to distinguish between facial expressions when the face in the photo is the face of their owner. This could be an indication that the distinction involved is a learned one. And there was simply more opportunity for the dogs to learn it from their own owner's face.

Rico

In 2002, the attention of a group of researchers in Leipzig was drawn to a dog said to be able to distinguish a variety of toys by name. Rico, a Border collie, appeared on the German television programme 'Wetten, dass …?' (Want to bet that …?). His owners bet that he would make no more than three mistakes in his search for the right toy. 'Rico, where is the BVB?' In his search for the BVB, Rico carefully rummaged through his toys. About 80 of them were laid out in a circle around his owner. It only took a short time before he had found the right one. The BVB, a black and yellow ball, was correctly identified by Rico and fetched for its owner. 'And where is the Joschka?' He was sent off again, and again it didn't take him long to make the right choice. This went on for minutes, and if Rico had had his way, it could have gone on for hours. Rico seemed to be able to keep these many different objects apart by name. But could he really? Or was there something else behind his behaviour?

Seeing Rico at work on television at that time first raised the following question: could it be that Rico is the 'Clever Hans' of the dog world? Clever Hans was a horse that lived at the beginning of the twentieth century and of whom it was said that it could do calculations. Its owner, Wilhelm von Osten, had trained the horse and found that it was able to solve simple arithmetic operations. The horse signalled the solution of 2 + 4, for example, by scraping the ground six times with its front hoof. People all over the world became aware of the abilities of this special horse, and very soon renowned scientists were also studying the phenomenon. Again and again, the horse was given problems in arithmetic, and again and again the horse was able to solve them. In 1904 it was then the psychologist Oskar Pfungst, who—having become aware of the horse—carried out studies with it.

His first suspicion was that Hans was not really solving the arithmetic problems but was instead making use of little hints from his environment. To investigate this question, the first step was to take these aids away from Hans. In this way, Mr Pfungst wanted to find out whether the horse could then continue to solve the problems posed. For this purpose, the scientist constructed a special set of blinkers that prevented Clever Hans from seeing his owner or any other person in the vicinity while the task was being performed. When Hans wore these blinkers, he could not solve any of the tasks. Now he scraped for as long as he wanted, no matter what arithmetic task was set. If the blinkers were taken off again, this changed, and the animal's abilities returned. The presence of people who knew the solution to the task was required for the horse to demonstrate its extraordinary arithmetic skills. Not only that, but Hans needed to be able to see the people surrounding him. A closer examination of this connection revealed that Hans had developed a remarkable sensitivity to cues that people around him were unconsciously transmitting. For example, some people turned their heads a little as soon as the horse reached the right solution with his hoof scratching. Others might hold their breath, and this was the signal for Hans to stop scraping. Unconsciously, therefore, it was the people who were solving the numerical problem and not the horse. The fact that the horse showed an extraordinary and in itself remarkable sensitivity in picking up on this was subsequently often forgotten. This case went down in the textbooks from then on as a negative example of the overestimation of a phenomenon that could be explained by simple means.

Was the Border collie Rico simply the Clever Hans of the dog world? Was he perhaps particularly sensitive to gestural cues that came from his owner and helped him to find the right item? Or was he really able to assign names to the numerous objects from memory? To find out, scientists from Leipzig carried out a simple test (Fig. 7.4). They posed the question whether Rico

Fig. 7.4 Some dogs can distinguish toys by name. They retrieve the correct items on command, for example: 'Fetch the duck!'

would still be able to distinguish the objects by name if no one was in the same room with him while he was looking through the toys. If Rico could not see anyone while he was searching, he would not be able to use any unconscious assistance. The researchers, thus, distributed Rico's toys in one of the several rooms in the flat in which he lived with his owners. The owner was in another room. When Rico ran into the room where the toys were, he could not see anyone while he searched. Likewise, no one could watch him while he searched.

It was said that Rico could distinguish among 200 toys. To test this, all the toys were randomly divided into groups of 10 toys each and questions were raised about two from each group. Rico's owner gave him the order to fetch one of the toys from the neighbouring room. Rico ran off, searched the flat, and came back with the right toy. By the end of this test, Rico had only made three mistakes, proving that he was not just a 'Clever Hans'. Rico was able to distinguish his two hundred different toys by name, even without receiving any assistance.

Rico was not an isolated case in the animal world. There are individuals of various animal species that have become real celebrities in science. The bonobo, Kanzi, for example, is one such case. Kanzi is able to assign terms to over 200 abstract symbols (so-called lexigrams). These include names for things like yoghurt, jam, coconut, and backpack. But Kanzi also knows terms like river, little finger, observation room, and floor. Among the lexigrams that the bonobo can use, there are also some that stand for verbs, such as carry, seek, and have. To communicate with his caregiver, Kanzi presses one of the lexigram symbols on a keyboard. This way he can communicate when he wants yoghurt, for example, or a coconut. He then presses the lexigram for 'yoghurt' and the lexigram for 'have'.

Another celebrity was the grey parrot, Alex. Alex was able to produce human vocalisations and thus to 'speak'. He used these human vocalisations in the right context. His 'vocabulary' ran to well over 100 terms. These included numbers, verbs—such as to leave or to want—and terms for various objects—such as paper, maize, or keys. From these, the parrot could even form word combinations such as 'wants maize' or 'wants to go away'.

Understanding Language

Does this mean that Rico, Alex, and Kanzi understand human language? That Rico, because he can use spoken words to distinguish toys, understood what humans were saying?

When we look at language comprehension, there are two levels of interest. First, how do people actually talk to dogs? Are there any special features in the way humans communicate with dogs? And secondly: what does the dog understand about what is being said.

In a study in eastern Kentucky, USA, it was found that people talk to dogs in much the same way as they talk to small children. People use so-called baby talk when interacting with their dogs. Among its characteristics, baby talk is spoken in a higher pitch. In addition, speech is simplified, and use is made of shorter sentences and repetitions.

Scientists conclude from the fact that people tend to speak to dogs in baby talk that it serves a function similar to that in children. Baby talk is meant to signal friendliness and affection. It serves to increase the bond with the other party and to create a basis of trust. This can be easily applied to communication between humans and dogs. A solid basis of trust is an important prerequisite for a successful relationship between humans and dogs. A further study by this research group demonstrates that baby talk has this function in interaction with dogs. It shows that people speak differently with other people's dogs than with their own. People praise other people's dogs more than their own and speak to them in longer sentences. People speak in a higher pitch and, on the whole, use more baby talk than when interacting with their own dog. The scientists assume that this serves to appear friendlier to the dog of another person and to placate it. It also served to establish a relationship of trust. It is a clear example that certain functions of baby talk can be applied to dogs.

But baby talk also has another important function. It is meant to increase the attention of the other party. A study from Sussex showed that dogs pay particular attention to photos of a person if a high-pitched voice is heard at the same time. This is also already the case in puppies and leads to the supposition that this attention to high-pitched human voices could be innate in dogs. Baby talk is particularly helpful when the other party needs to be taught something, for instance, a new term. In this context, it is not surprising that it is used especially when the other party is limited in its ability to understand what is being said. This is just as true when dealing with dogs as it is when dealing with small children. The tendency of people to use baby talk in these situations is presumably innate. It has evolved to help children learn language.

This is consistent with the fact that women use more baby talk than men when dealing with children. After all, in many cultures it was the women who took care of the children. It was probably also women who took care of the dogs (see Chap. 2). Perhaps this is why women use much more baby talk than men when dealing with dogs. Baby talk helps children acquire language. But should it also fulfil this function for dogs? In this context, it is interesting to take a closer look at the differences between baby talk addressed to children and baby talk addressed to dogs. Children are spoken to in explanatory terms. In relatively long sentences and without many repeated words. This corresponds to the pattern that one would expect if one assumes that baby talk is supposed to help with language acquisition. In contrast, dogs are spoken to in short sentences.

Moreover, people use more of a commanding tone than an explanatory one. Words or whole sentences are also often repeated. Thus, help with language acquisition is not a direct function that baby talk is supposed to serve when dealing with dogs.

We speak to dogs in a similar way as we do to children, but what part of it reaches the dog? Does the dog understand what we say? And what part of what is said matters to the dog? As early as the 1930s, researchers in Leipzig dealt with this question. For this purpose, a dog was trained to sit down dependably when given the command 'Sit'. Then the command was modified and 'Sit' became 'Fitz' or 'Sense'. It turned out that the tested dog's response to all modified commands was significantly worse. Scientists from London followed the same idea. However, they examined a large group of dogs. Here they also found that the dog's response to modified forms of commands for 'Sit' or 'Come' was worse than to the original command. For example, if 'Sit' became 'Chit', the dogs sat down significantly less often. Other factors also played a role here. Two circumstances were particularly important for the dogs. They had to be able to hear the original voice of their owners. If the voice was played back from a tape, the dogs ran into difficulties. Interestingly, it became particularly problematic if there was no opportunity for the dogs to see their owner's eyes in this situation. Thus, if the owner's eyes were covered with sunglasses and his voice was coming from a tape, the dogs did not comply with the command. Perhaps it is the case that when the owner's voice comes from a tape recorder, the dogs can at least orient themselves according to the owner's facial expressions. Therefore, they run into particular difficulties when deprived of this opportunity as well. It has already been discussed in Chap. 3 how attentively dogs watch their owner's eyes. This shows once again how important an owner's facial expressions are for dogs in situations of uncertainty. If the owner's voice does not come from a tape and the circumstance is not uncertain, then the dogs do not care if they can see the owner's eyes. They follow the command regardless whether their owner is wearing sunglasses or not.

Small changes in the commands have an impact on a dog's response. However, when the external circumstances are right, dogs seem to listen very carefully. This explains why Rico was even able to distinguish between words like 'Yoschka' and 'Oscar'. However, he sometimes had problems when words were too similar. For example, he sometimes confused terms like 'Handy' [= 'mobile phone' in German] and 'Wendy'. However, this rarely happened, and Rico was able to distinguish a remarkably large number of toys by name.

Learning New Terms

One question that particularly interested the researchers in Leipzig was how Rico was able to learn new terms for new objects. The researchers wanted to know if Rico was able to learn terms in a way in which it had previously been assumed only humans could. A special feature of language acquisition is that children learn at an impressive speed. Children at the age of two learn up to ten new words a day.

When Rico learnt new names, his owners said, it was enough to repeat the term once or twice. You held up the object in front of him, said the new name, and he was already able to distinguish the new toy from the old one. When children learn new terms, they are not only able to memorise new ones quickly, they also use special learning mechanisms that make it easier for them to acquire language quickly. For example, they can grasp new terms *indirectly*. Let's assume a child already knows the term 'blue' and is shown a blue and an olive-green object. Now you ask the child to fetch the 'olive-green' object and not the blue one. Three-year-old children are able to understand that the term 'olive green' must stand for the green object and not for the blue one. Thus, they are able to use a process of exclusion in order to assign the new term to the unknown colour. But that's not all. Using this process of exclusion, they also *learn* that this colour is probably called 'olive green'. This method enables children to grasp many new terms very quickly.

The Leipzig scientists who were working with Rico now asked themselves whether it was possible that Rico used a similar mechanism to learn new terms (Fig. 7.5). They wanted to investigate how quickly and under what conditions Rico learnt the names for his toys. They also wanted to find out how flexibly Rico was able to make use of his 'knowledge'. Thus, they again put a few of Rico's familiar toys in a room, but now they had added in a toy that Rico had never seen before. First, Rico was asked by his owner to fetch one or two of the familiar toys. In this way, the researchers wanted to rule out the possibility that Rico would fetch the new toy first simply for the joy of it. Then it would not be clear whether he really understood that the new word stood for the new object.

Fig. 7.5 Rico learnt the new term by a process of exclusion. ('Frühling' is the German word for 'spring' [the season])

So after Rico had fetched one or two of the familiar objects, there was another search command. Now the owner used a word Rico had never heard before in connection with toys, such as 'spring'. Confronted with this new name, Rico chose the new toy from among all the others. To make sure that Rico's behaviour was not just a one-time coincidence, this experiment was repeated ten times. Each time, a different strange toy was mixed in with a lot of familiar toys. Rico fetched the new toy seven times out of ten when the new name was used. Thus, it was no coincidence. Rico was able to assign a new name to a new toy by means of a process of exclusion.

The question that followed from this was whether Rico had also learnt the new names for the new toys via this exclusion process. To check this, Rico was initially deprived of the newly introduced toys for 4 weeks. After this time, another test was supposed to show whether Rico still remembered the names for the new toys. For this purpose, different toys were again hidden in one of the rooms in the flat. Four toys that were known to Rico, four toys that were completely unknown to Rico, and the one toy whose name Rico had learnt 4 weeks ago by a process of exclusion. Even after these 4 weeks, Rico still correctly recognised three of the six toys presented. Statistically, however, this is a random success rate. So nothing learnt?

If Rico was given the task described above just 10 min after the first assignment of the new name, he correctly recognised four of the six toys presented. Statistically, this is no longer a coincidence. So Rico was not only assigning new terms to new objects, he was also learning these new terms as he did so.

This kind of learning is called fast mapping. As mentioned earlier, this was thought to be one of the learning mechanisms that only humans use when learning new terms. Since Rico was also able to learn new terms in this way, this cannot be true. So there is an unexpected commonality between dogs and humans. But does this mean that Rico understood everything that was said? That even though he couldn't speak, he had at least learnt to understand? It's certainly not that simple. Rico seemed to have understood a rule. The rule held that toys have names. This rule allowed him to use the exclusion procedure. However, Rico's owners could still talk freely about the vet visit planned for the next day without their dog cowering anxiously under the nearest table. Even a conversation about the treats hidden in the living room could not induce Rico to go in search of them immediately.

Questioning Rico's Abilities

There are scientists who study children's word learning and who say that Rico's abilities cannot really be compared with it. For one thing, there is the question of whether it always had to be the owner who gave the search commands. This could be an indication that Rico's understanding of the situation was not very flexible. That it always had to be the same voice, i.e., the same stimulus, for him to be able to perform. However, this was not the case. Anyone could have asked Rico. You could have asked too, and he would have retrieved the right object. The owner could even disguise her voice, and the dog was still able to distinguish the toys. In this respect, Rico was very flexible in his abilities.

A further point of debate was whether Rico really linked the new name to the new object using the exclusion procedure. Critics say that it could be that Rico did not really understand that 'spring' stood for the new object. That is, that he did not fetch the new object out of an understanding of the relationship between the new term and the new object. It could be that he simply couldn't wait to finally get the new thing. In this case, the new term might simply be like a letting go command. Finally he was allowed to do what he wanted, he was allowed to get the new toy. But if this had been the case, how in fact did Rico learn the new terms? 'Spring' was not repeated as a word as soon as Rico got to his owner with the toy. The only moment he was able to associate the new object with the new name was the moment he chose it from among the others. But this also meant that he did not simply respond to a 'let go' command. Rico must have had an idea which toy was 'spring'. He knew all the others, it had to be as the unknown one.

A final point of debate was whether Rico really understood the command 'Fetch the object'. This sentence contains the verb 'fetch' and the name of the object referenced by the verb. At this point in time, it is not clear whether

Rico was really able to understand the meaning of these words and relate them to each other. For example, would he also have been able to pull an object? Or to push one? Would he have been able to understand a sentence like 'pull the pony and push the parrot' and perform these actions in sequence?

This is where a study recently conducted in the US with another super talent, Chaser, comes in handy. Chaser is a 4-year-old female Border collie who, under controlled conditions, proved that she could distinguish 1022 objects by name. That in itself is extraordinary. However, Chaser has not only mastered the names of objects, she can also do various verbs. She is familiar with the command 'Take'—whereupon she takes the object into her mouth. But she is also familiar with the command 'Paw'—whereupon she touches the object with her paw. She is equally familiar with the command 'Nose', whereupon she touches the object with her nose. She is able to perform different actions (i.e., 'take', 'paw', or 'nose') with different objects. By doing so, she shows that she can interpret both parts, i.e., the verb and the name of the object as different pieces of information. She understands that the respective verb *refers* to the object.

An Exceptional Talent?

Chaser teaches us that Rico is not an exceptional talent, not a miraculous Einstein of the dog world. Similarly, a study conducted by scientists in Leipzig shows that there are at least two other dogs besides Rico and Chaser that can distinguish up to 300 toys from each other. To this end, they were tested under the same conditions as Rico and Chaser. Just like these two, they showed corresponding abilities. This makes it clear that other dogs, although not all, also have the talent to distinguish a large number of objects by name. The ability also seems to be independent of the dog's breed to a certain extent. For example, the scientific literature describes a female Yorkshire Terrier that could distinguish two hundred objects by name. Interestingly, however, all other dogs in which this ability has been scientifically studied have been Border collies.

Border collies are considered by dog connoisseurs to be particularly smart dogs. And although this appraisal is stubbornly held on to, there are actually no systematic studies to date that prove this in any way. So could the fact that the dogs with the described abilities have been exclusively Border collies be proof of the extraordinary smartness of this breed? Or is there perhaps another explanation for the connection between breed and the abilities described here?

Border collies are herding dogs, specially bred to lead a flock to their shepherd. It can be assumed that they therefore show an innate 'fetching' behaviour. Border collies therefore have an innate urge to fetch, and thus to retrieve, from birth. If we now look at the task that Rico and the other dogs do so well, it is clear that dogs that do not retrieve are ill-suited for this task. Border collies or, more generally, dogs that like to retrieve objects are, in contrast, ideally equipped. If a dachshund were to fail at this task, this should not be taken to mean that it does not possess this ability in principle. And if a Rhodesian ridgeback cannot be motivated to retrieve, how can the basic procedure of this study be conveyed to it?

Besides the fact that Border collies show innate retrieving behaviour, they are also considered to be highly motivated working dogs. This makes Border collies a demanding breed that is only suitable as a family dog to a limited extent. These dogs need to be nurtured in a special way. They can only be integrated into normal family life if the owners are prepared to meet these high demands. This is especially true if the dogs stem from a breeding line specifically bred for work. Border collies are therefore dogs that love to retrieve and are also highly motivated. So when considering the abilities of a particular dog in a particular situation, the breed of the dog and thus its innate behavioural patterns inevitably play a part.

Another question that interested Leipzig scientists was how flexible the communicative understanding of these Border collies is. It is evident that these dogs can distinguish the objects presented to them by their names and thus by means of an audible signal. But what happens when the dogs are presented with a visual signal instead of an audible one, for example, with a symbol (Fig. 7.6). Will the dogs still be able to find the right object? To find out, the scientists confronted three Border collies with the following situation: Eight familiar objects were arranged for the dogs in a room. Since the dogs were kept waiting in a neighbouring room, they did not know which objects were where. They were now given a visual signal by their owner as to which object they should retrieve. Instead of saying the name of the object out loud, the owners held an exact copy of the object to be retrieved in front of the dog's eyes and said, 'Fetch this!' All three Border collies knew immediately what was expected of them. They identified the correct object in the neighbouring room and fetched it for their owner. The scientists then made the situation more difficult. Instead of an exact copy of the object, they had the owner present merely a much smaller miniature version of it. This was to show whether the dogs also knew what to do with a more abstract cue. Again, the Border collies had no difficulty in identifying the requested object. In another test variant, the owners simply presented a full-size photo of the

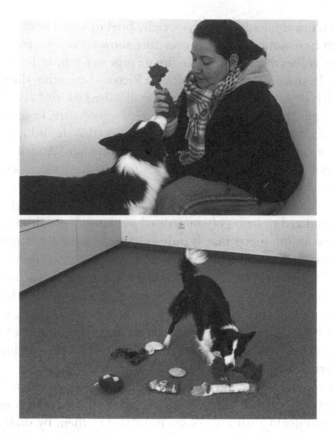

Fig. 7.6 The language-trained Border collies see a copy of the familiar toy and are given the new command: 'Fetch this!' Thereupon they fetch the original

object they were looking for. And suddenly only one of the three Border collies succeeded. The other two did not know what to do with this cue and were unable to identify the correct item.

One question that naturally also arises is why so few dogs with these special abilities have been found so far. Is it perhaps a case of particularly talented dogs or particularly effective training on the part of the owners? All dogs that are able to distinguish this large number of toys had already been introduced to this game as young dogs. Although there is no reason to believe that a dog cannot learn this later in life, the early introduction to the task has certainly been beneficial. In some cases, this arose, as it were, out of a situation of need. Due to injuries, two of the dogs were not allowed to run free or move around much. Keeping a young dog that cannot be put on a lead occupied so that it is tired by evening is certainly a challenge. This forces the owners of such a dog to come up with something. And this is where a game of hide-and-seek that can be played indoors comes in handy. You may be familiar with this. Objects are hidden in the flat and the dog is called upon to look for them. Your dog will systematically comb through the flat, always looking for the desired object. Teaching the dog the names of toys and letting it distinguish among them will keep it sufficiently mentally active and, in an emergency, replace its walk.

So is it the combination of these different factors that has led to the special abilities of these dogs? Innate retrieving behaviour, a high level of motivation, and early mental nurturing? Certainly all these factors play an important role, but it doesn't seem to be that simple. There are reports of dog owners who own two Border collies. Both have grown up in the same way, both have been trained and nurtured in the same way, but only one of the two dogs shows the ability to distinguish objects by name. Add to this the fact that the owners of all three tested dogs report that it was the dogs who started this game, so to speak. Quite automatically, one says to the dog 'Fetch the horse' and not simply 'Fetch the toy'. So you give the toys names if you want to distinguish among several of them. After a while, the owners of the three tested dogs noticed that their animals really began to distinguish among the toys. They noticed that when the dog was called on to fetch the horse, it always fetched the horse and not the ball or the crocodile. Distinguishing among the toys was only trained to a limited extent. So is this a special talent after all? Something that not every dog necessarily has, no matter how much it is fostered? These and many other questions will certainly be the subject of future research.

Summary

So what have we learnt about communication between dogs and humans? We now know that dogs listen quite carefully when they are given a command. The way commands are spoken is therefore not entirely unimportant. No dog owner should be ashamed if he or she speaks to his/her dog in baby talk. For, first of all, it is difficult for us to suppress this urge anyway. And secondly, baby talk also serves its purpose with dogs.

Furthermore, we know that dogs have a wide range of different kinds of vocalisation at their disposal, which they have probably only developed in the course of living together with humans. Even people who do not have a dog can assign different emotional states to such sounds. And probably such emotional states are the only things we are dealing with here.

In addition, we have seen that dogs pay attention to people's facial expressions and can distinguish between them. By the same token, humans can also read the facial expressions of dogs.

Finally, we have learnt that there are some great talents in the dog world. They possess, for instance, the ability to distinguish a large number of different terms from one another.

But even dog owners who have a dog with above-average talent don't have to worry about what they talk about within earshot of their dogs. For to say that this ability has anything to do with a true *understanding* of human language would be pure speculation at this time.

8

What Do Dogs Know About Their Environment?

A scene in the foyer of the Institute: We have just returned from a walk. Now my colleague would like to pick up a coffee in the cafeteria. Her male German shepherd Benny trails behind his owner. But dogs are not allowed in the cafeteria. I will just take him with me to the office and lead him in the direction of the lift. It's in the middle of the foyer. From there you can look through the large glass windows into the cafeteria. Benny obediently comes with me into the lift. But as soon as he is inside, he turns around. He pulls on the lead—in the direction of his owner, whom he can still see. The lift door closes and opens again seconds later on the first floor. No owner to be seen any more. Nonetheless, Benny seems to be—unsurprised. He no longer pulls on the lead and lets himself be led into the office without any trouble. Of course, you will say. He knows that he went up one floor in the lift. But has he really understood that?

In the previous chapters of this book, the focus was on what dogs understand about humans. We talked about whether they learn from humans and how they respond to words and gestures. But all the situations described were social. That means that a person, or at least a fellow dog, always played a role.

Now let us take up the question of what a dog understands about its environment. Can it understand simple physical relationships? Does it know, for example, that objects that are dropped always land on the ground? Does it 'wonder' when something inexplicable happens? How does it find its way around its environment? Can it distinguish different quantities from one another?

But what interests us most is how flexible these abilities are. The question is whether dogs can adapt to new situations. Our dogs have shown themselves

J. Bräuer, J. Kaminski, *What Dogs Know*, https://doi.org/10.1007/978-3-030-89533-4_8

to be very flexible when living together with humans. For example, we have seen that dogs are not only sensitive to human attentiveness when they beg, but also when they snap at forbidden food or fetch a person a toy. Are dogs just as flexible when it comes to their inanimate environment?

Object Permanence

First, let's return to our male German shepherd, Benny. Maybe he knows exactly how a lift works. But that is relatively unlikely. If he doesn't know, it's strange that he wasn't 'surprised' when the lift door opened and his owner was suddenly out of sight. Why wasn't Benny surprised? Had he forgotten in such a short time that his owner was standing there when the lift door closed? Doesn't he know that someone or something continues to exist, even if you can't see it? Out of sight, out of mind?

Of course not, you will say, otherwise it wouldn't be any fun to hide your pet's favourite toy. For it looks for the ball once you have made it disappear. Some dogs even retrieve the stick from the bushes from yesterday's walk. And my dog Mora never seems to forget the existence of a compost heap from which she once successfully retrieved a bone.

So it looks like dogs have an understanding of object permanence. That is, they understand that someone or something continues to exist even if you cannot perceive this at the moment. Experiments by different research groups have confirmed this. How do you test object permanence? As always, it involves toys or food. In addition, several barriers are placed in a row next to each other. The dog sits opposite them so that it cannot see behind them. Now a ball is hidden behind a barrier in front of the eyes of the test animal. The question is very simple: Where will the dog look? The exercise is not difficult for it; it goes to the right barrier to find its ball again. In this way, it acts like an 8-month-old baby. If it sees you hiding a toy under a blanket, it will lift the blanket to get the toy again. So like children at this age, dogs don't forget the toy, but look for it in a purposeful way.

But perhaps there is another explanation for the dogs' behaviour. Maybe they only went to the right barrier so purposefully because they smelled their ball? Scientists from Quebec, Canada, wanted to rule this out with certainty and resorted to a rather unique method. They used four identical rubber toys. All four barriers were sprayed with strongly scented rose water just like the toys. They also placed three of the four toys behind the barriers before the test began. Now, as before, one toy was hidden in plain sight of the dog. In other words, now all four barriers smelled the same: of rubber toys and rose water.

Here, too, the dogs ran to the right place. In this way, they proved that they could solve the problem not with the help of their nose, but with the help of their head.

The Ball in the Container

But what happens when the task becomes a bit more difficult? When the dogs cannot directly observe how the toy is hidden? In these experiments, in addition to the ball and the barriers, a small container also plays a role (Fig. 8.1). Now the procedure that the dogs see is a bit more complicated: The ball is put into the container, which then disappears behind one of the barriers. When the container comes out again, the dogs are shown that it is empty. Since the ball can hardly have disappeared into thin air, it must now be behind that barrier. In order to understand this course of events, you have to visualise in *your head* what you cannot see. Not only do you have to *imagine* the ball being taken out of the container and put behind the barrier. You also have to *remember* where it was. Are the dogs able to do this? Various research groups have carried out these experiments. Usually, the dogs run to the correct barrier—behind which the container with the ball had previously disappeared.

In Quebec, dogs were faced with both the easier and the harder task. The toy was either hidden right in front of the test animals. Or it went first into the container and then behind the barrier. It is certainly not particularly

Fig. 8.1   The ball is placed in the container in front of the dog. The person goes behind one of the barriers with the container and then shows the dog that the container is now empty. Where is the ball?

surprising that the dogs solved the difficult problem with the container but performed worse than with the easy one. In addition, the researchers found that the dogs then used their nose more. They sniffed the ground, the barriers, and the container. Even if this did not help them, they obviously were trying to get additional information when they were unable to solve the task so easily.

But what made the task more difficult for the dogs? Was it perhaps their memory? Can't the dogs remember where the toy is if they don't directly see where it is hidden? The scientists found a simple way to answer this question. Under one test condition, they let the dogs loose immediately after hiding the toy. In the other case, the test animals had to wait 10 or 20 s before they were allowed to look for their toy. The results were unequivocal: the dogs always performed equally well, regardless of whether they had to wait or not. This means that they were able to remember where the container or the ball had gone. The dogs are able to solve the problem and go to the right barrier. However, scientists are still debating whether they really understand the course of events. Some assume that the dogs are guided by simple rules. For example, they look for the ball according to the rule: 'Go to the barrier where the container last was'. Or: 'Go to the barrier next to where the container is'. In any case, the dogs are highly interested in the container in which the ball disappeared.

There is something else that might have influenced the dogs in this experiment. That is the person himself. We saw in Chap. 6 how good dogs are at interpreting human gestures as a cue to hidden food. We also know that objects become interesting for dogs simply because a person has taken an interest in them. In our experiment, the ball must somehow get into the container, and the container must then get behind the barrier. The scientists always tried to make sure that the person did not play a big part in this. For example, a wooden handle was attached to the top of the container. In this way, the person never touched the container directly, but moved it with the help of the handle. The toys were also usually attached to a string so that they did not need to be touched. Nevertheless, we cannot completely rule out the possibility that the person's behaviour influenced the dogs in their search.

So we still don't know exactly what goes on in the dogs' minds when they are confronted with this problem. Here again, a comparison with children can help us. In Budapest, 4- to 6-year-old children were tested to this end. The experiment went a little differently here. The toy was again put into a container. This time, however, the container disappeared behind all three barriers one after the other, only to reappear empty. The children could only guess where the toy was hidden. But they could conclude, however, that it was hidden behind one of the barriers. After all, it could not be anywhere else. Now the children were called upon to find the toy again. Sometimes they guessed the right place immediately. But we are interested in the test runs in which they got it wrong. If they did not find the toy behind the first barrier, they quickly ran to the second, and even more quickly to the third. Clearly, the children understood the logic of the game perfectly: if the toy could not be found behind the first and the second barrier—then they were quite sure: it had to be behind the third! If they had been asked where they would find the toy after having visited the second barrier without any luck, they would certainly have answered: 'There, it *must* be behind the third barrier.'

Unfortunately, dogs are not able to answer such questions. But you can test them in the same setup. That's what the scientists in Budapest did. The dogs also looked for the ball. If they did not find it behind the first barrier, they ran to the second and the third. The decisive difference to the children was that they did not become *faster*, but *slower*. Every negative experience—namely that the ball could not be found behind a barrier—obviously lowered their incentive to keep looking. It is not likely that they lost interest in the toy so quickly. My dog Nana would spend hours looking for a ball that had rolled into the bushes if she was allowed to. The dogs in the trial certainly wanted to find the ball too. Because when they found it behind the third barrier, they played with it. There was obviously another reason why they became slower and slower in their search. They were discouraged by not finding the ball. And unlike the children, they were unable to conclude that it *had to be* behind the third barrier. In other words, they hadn't quite grasped the logic of the game.

So what does all this mean? Dogs do understand that their toy continues to exist when they no longer see it. They look for it in the place where it disappeared. If the toy is first put into a container and then disappears behind a barrier, they also usually look in the right place. But they obviously can't imagine how the ball is taken out of the container and put behind the barrier.

Shell Games

Understanding in their minds how a hidden object is moved is something that obviously causes difficulties for dogs. This is also shown by so-called 'shell games'. You too have certainly witnessed the sleight of hand involved when two nimble hands hide a coin under one of three matchboxes and then move it with lightning speed. If this is done very quickly, you will hardly be able to guess at the end where the coin is hidden. If, however, the boxes are moved so slowly that you can follow each individual move, you will certainly have no difficulty finding the coin again. Our closest relatives, the great apes, can also solve this problem. In the experiment, chimpanzees and orangutans were tested. Upside-down cups were used in place of matchboxes. And the reward, of course, was not a coin, but pieces of orange or grapefruit. The ape sat across from a person. There were two cups on a wooden board between them. These were turned upside down and stood next to each other about half a metre apart. Now the piece of orange was put under one cup in full view of the apes. Then the cups were swapped, with the human pushing them along the board. Now the ape was supposed to point to a cup, which was then lifted. This was lifted. If the food was there, they were allowed to eat it. But if they were wrong, there was no orange for them, and the next round began. The apes chose correctly straightaway. Even if the cups were moved a second time, returning them to their original position, the apes again found the piece of food. Chimpanzees and orangutans are apparently able to *imagine* how the piece of orange is shifted along with the cup.

And how do dogs fare? You can easily recreate this experiment (Fig. 8.2). Just as with the apes, place two upside-down cups in front of your dog. You hide a piece of food or his favourite toy in one of them in full view of your pet and then switch the two cups around. You are going to be a bit disappointed. Your dog will presumably always go to the wrong cup. It will look for the food where it was hidden. It obviously can't understand that the piece of food is

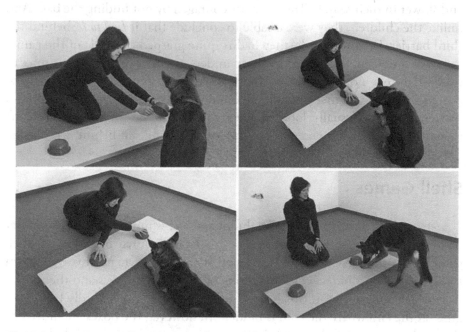

Fig. 8.2 The German shepherd sees how the food is hidden and moved with the cup. But the dog looks for the food from where it disappeared

moved along with the cup. Even if you simplify the task a little, your dog will make this mistake. Two different looking cups should make things much easier. If the shell player uses a red, green, and blue matchbox, you will always choose correctly. You understand that the coin is moved along with the box. You just need to remember that the coin has gone into the red matchbox. Now even lightning-fast, back-and-forth movement can no longer confuse you. You choose the red box. Your dog will not be helped by cups that look different. This is certainly not because it is unable to distinguish the colour and shape of the cups. It is certainly not because it is unable to remember where the food has disappeared to. It's because the dog doesn't understand the whole process of moving it around. Perhaps you can take a little comfort from this: cats don't do any better at this exercise.

There is only one condition under which dogs—and cats as well—can solve such a problem. Namely, when the cups are moved sideways in one direction. Let's assume that the left cup is full and both cups are moved to the right. So now there is no cup on the left, the filled one is in the middle, and the empty one is on the right. This means that there is now nothing left where the food disappeared from. Now the dogs choose correctly. But this presumably has nothing to do with the fact that they understand that the food is moved along with the cup. They are simply searching in the cup that is closest to the place from where the food disappeared. Presumably, they also use the direction of movement during the shift as a cue. Thus, again, the dogs are not using their understanding to solve the task, but rather a certain strategy, a strategy that we will continue to come across in the following.

But why do dogs obviously not *understand* such hiding processes? Why can't they understand in their minds how a hidden object is moved? Yet they do very well in other tests, sometimes even better than the great apes. The answer is quite simple: because they don't actually need this kind of understanding. After all, it is a very sensible strategy to always search in the place where the toy has disappeared. In most cases, this will prove successful for the dogs. After all, hidden objects are rarely moved. For example, you put your pet's ball in the cupboard when you come back from a walk. The cupboard will certainly not be moved. The ball will stay there until you take it out again for the next walk. And that is something your dog also understands.

The Magic Cup

Now the question is whether the dogs also remember exactly what has been hidden. For example, if you hide food in a cup with a lid, do they simply remember that there was something to eat? Or do they expect exactly the food that was hidden? Researchers from Leipzig have tested this in dogs and great apes (Fig. 8.3). The procedure was quite simple. A piece of food was put into a cup in full view of the animal. The lid was closed and immediately opened up again. However, the cup was designed with a double bottom so that another piece of food could be revealed when the cup was reopened.

The researchers wanted to know if the animals would be amazed if there was suddenly different food in the cup. A tasty piece of sausage—for the dog—or a grape—for the chimpanzee—disappeared into the cup. The cup was opened again, but suddenly there was only a boring piece of bread or carrot. But how does a dog or a chimpanzee look in amazement? Of course, it doesn't open its eyes wide or say. 'Oh!'. Nevertheless, it can be gleaned from its behaviour. Both animals ate the boring food, but then searched for the food that had disappeared. The chimpanzees bent forward to look closely into the cup, while the dogs smelled the cup and lid. They did not do this if the same piece of food was in the cup after closing the lid. So they remembered very well exactly what had been hidden. Even in the opposite case, when boring food had been put into the cup and then suddenly a tasty piece of sausage or a grape appeared, the animals searched for the original food. They thus

Fig. 8.3 The Labrador sees how a sausage is hidden in the magic cup. The magic cup is then closed. When the magic cup is reopened, however, there is only a piece of bread in it. The Labrador searches for the missing sausage with his nose

proved: not only do they remember that something edible had been hidden in the cup, but they also remember exactly what kind of food it was.

Now you may think: have the researchers disregarded the dog's good nose? After all, couldn't it have smelled the food still present in the false bottom? But the researchers used a little trick to check this. In some test runs, a piece of sausage was hidden and thus wandered into the false bottom when the cup was closed. When the cup was opened, the dogs found an identical-looking piece of sausage. This time they did not sniff the cup, they ate the sausage and went away. Their sniffing behaviour thus had nothing to do with the sausage in the cup—it was still hidden in the false bottom—but with their expectation. And that expectation had been met: Sausage hidden, sausage back. Yummy!

What's Where?

Dogs expect to find exactly the same food that was hidden in the cup. Now scientists wanted to know whether dogs can also remember what was hidden *where*. True masters of this kind of exercise are, by the way, birds that store food, such as the scrub jay. They are known to remember not only *what* food was hidden *where*, but also *when*. Suppose tasty worms and less tasty nuts are hidden. If the birds are allowed to go to the hiding places after a short time, they will get the favoured worms. But if they are only allowed to fetch the food from its hiding places after 5 days, they then go for the nuts. Apparently, the scrub jays conclude that the worms have spoiled by then! So they remember *what* was hidden, *where, and when*.

In the case of the dogs, the first question was whether they remembered *what was* hidden *where*. For this purpose, Rico was tested, as well as another Border collie that could also distinguish objects by name. Twelve objects that the dogs were familiar with were distributed across two rooms. The test ran like a normal search game. The owner sat in a third room and called Rico to her. She asked him to fetch one of the toys: 'Fetch the frog!'. In the first run, Rico did not yet know which of his many objects were lying there nor how they were distributed. So he ran through both rooms until he found the desired object and then fetched it for his owner. He had already proved that he could do this in other tests. In the following test runs, the owner asked for all 12 toys one by one. The scientists wanted to know *how* Rico would search for them. And the amazing thing was: from now on Rico always searched in the right room! He had gone through the two rooms a single time and had immediately memorised what was hidden where! This is a feat that many an unfocused adult would have difficulty with!

Now the question arises again whether this is typical 'Einstein' Rico. The other Border collie tested was less determined in its search. However, it is quite possible that other dogs are also good at remembering what has been hidden where. But you can't test them in the situation described, as long as they are unable to distinguish objects by name.

What Information Is Important?

In the following, we will look at the rules dogs use to search when they cannot know where the reward is hidden. This study was conducted in Quebec. It was about toys and barriers. The dog saw its ball being put behind a barrier. Then a curtain was drawn in front of it. Thus, the tested animal could not observe what then happened. The barrier was pushed to one side. A second barrier was placed on the other side. In the middle, where the ball had disappeared from, there was now nothing at all. At this point, the dog was allowed to look for its toy. If you were faced with the same situation, you wouldn't know either where to look at first. You would simply guess where the ball might be. But what would you do if—like the dogs—you were not told where the toy was for 20 test runs? Then you would probably ask yourself according to which rule the ball is hidden. Such rules did indeed exist. One half of the dogs could follow spatial cues. For example, the barrier with the toy was always moved to the right. To find the toy, the dogs simply had to search behind the right barrier. They learnt this very quickly.

The other half of the dogs were supposed to follow a different rule. This time, the barriers were distinct from each other. One was red with yellow squares on it and was placed vertically. The other was horizontal and had black circles on a white background. The ball was now always to be found behind the red barrier. This time, the dogs were thus able to use the characteristics of the barrier: its colour, its pattern, and its shape. However, they obviously found this difficult. In any case, they did not learn to always search behind the red barrier within the 20 test runs.

Spatial cues obviously play a greater role for dogs than the characteristics of the barrier. The rule: 'Always search behind the red barrier!' was much more difficult for them to follow than the rule: 'Always search behind the right barrier!' These results are also consistently confirmed in other experiments. For example, in the object choice test described in Chap. 6. In these tests, food is always hidden in one of two cups. And the dogs don't know in which. Now they get a hint where the food is and have to decide. If the dogs can't make any sense of this clue, then they often favour one side. Just like in the experiment with the barriers, they then follow a spatial rule. For instance: 'Always go right!'. Or 'Always go left!' This strategy, in turn, often leads to success, because the food is hidden on one side in half of the cases!

Orientation in Space

A highly important ability that, in principle, every animal needs, is to orientate itself in the world. Accordingly, scientists have posed the question of how dogs do this. This was something they also tested when they allowed the animals to search for their toys behind barriers. Here, for instance, the question was: do dogs orient themselves *egocentrically* or *allocentrically*? Now what does this mean? Say, for example, you are looking for Forest Road in any given town. You have two options. One, you can ask a passer-by at the side of the road. He or she will explain to you: 'Go straight ahead, then turn right. After the next set of traffic lights, keep half-left. The railway station is then on your right, and Forest Road begins on your left.' In other words, the passer-by describes the way to you by starting from your vantage point. He or she sets out your immediate surroundings in relation to your current position. This is termed an *egocentric* orientation. You are, so to speak, the centre of attention.

You also have a second option, however, for finding Forest Road. You can use a map of the city. Here you see the whole thing from above. You see that Forest Road lies to the north of the railway station. In this case, your position does not matter. You relate things to one another. Here you are orienting yourself *allocentrically*.

Thus, we as humans can do both. Depending on the situation, we either refer to ourselves: 'Forest Road is on my left', or to things in relation to one another: 'Forest Road is north of the station'.

But how do dogs behave? They obviously don't use city maps. Do they therefore always take themselves as their starting point when they orientate themselves? To find out, scientists from Quebec, Canada, developed a somewhat complicated test (Fig. 8.4). A toy disappeared from the dog's view into one of three boxes. We will refer to it as the target box. A curtain now prevented the dog from seeing what happened next. The boxes were moved. There were several options. For example, the target box was in the middle. All three boxes were now shifted to the right in parallel. But the dog could not know this when it went to search. The target box was still in the middle of the three boxes. But in relation to the dog, it was now one position further to the

Fig. 8.4 The poodle sees how the bone is hidden in the middle box. All three boxes are moved. The poodle now looks for the bone in the right box, taking itself as its starting point and orienting itself egocentrically. If it oriented itself allocentrically, it would choose the middle box

right. If the dogs related the boxes to one another, then they should go to the middle box. If they took themselves as their starting point, then they should choose the box on the left. Which is what they did. So they oriented themselves egocentrically. They remembered that the toy had just disappeared in front of them. Just like we do when we realise, 'Forest Road is on my left.'

Further tests have shown, however, that dogs are also able to orient themselves *allocentrically*. That is, if there is no other option, they can also relate objects to one another. In this case, they remember that the target box is the middle one of the three. Which would therefore be comparable to our assessment: 'Forest Road is to the north of the station.' According to this, when they have to orientate themselves in space, dogs behave very flexibly. Like humans, they can use two different ways to reach an unknown destination.

Can Dogs Count?

So far, we have looked at what dogs understand about hidden toys and how they search for them. Now we will look at whether they also pay attention to *how many* things are at play.

Scientists from Michigan have asked themselves whether dogs can recognise differences in quantity (Fig. 8.5). This can easily be tested in your own dog. It just needs to be a bit of a glutton. Let it choose between two small heaps of dog pellets. On one pile there should be five pellets and on the other only two. We assume your dog would rather eat more than less. Does it go for the five pellets? Presumably yes. Dogs are good at distinguishing between one and five pellets. The larger the difference, the better they are at it, just like we are and like apes are too. In other words, they are better at distinguishing between two and five than between three and four.

Fig. 8.5 Dogs can distinguish between five pellets and two pellets

Some dogs are even able to do this in their heads. Both sets of pellets are in small bowls with lids. When the test begins, both heaps of pellets are covered. Now one of the lids is lifted so that one pile is visible to the dog. Then this lid is closed again. Then the other lid is lifted and closed again. Now the dogs have a choice. The crucial thing here is that they have to solve the problem in their heads, as they never see both quantities at the same time. So they have to remember the number of pellets in each bowl and compare them in their heads. And this they can do.

Now Italian and American researchers wanted to know whether dogs choose food more according to the number of pieces, i.e., its quantity, or according to its size, i.e., its volume. For example, the animals had to choose between four small and two very large food pellets. The two large pellets were larger in total volume than the four small pellets. Thus, for a hungry dog, they were the better choice. It turned out that dogs did not regard the number of food pieces as decisive, but the volume. They usually chose the two large pellets. That makes sense, after all, if the aim is to eat as much as possible.

Interestingly, the dogs can be easily distracted from their 'calculating skills'. Namely by their owners. In this test, two different portions of equal-sized pellets are shown. If the owner now bends over the smaller portion and enthusiastically praises the tasty food, the dogs suddenly choose this smaller portion. They trust, so to speak, their owner more than their own perception. That such a social cue is more important to the dogs than what they see themselves is something we will encounter frequently.

Gravity

Let us continue to pursue the question of what dogs know about their inanimate environment. Do they have expectations about certain physical principles? For example, gravity? Do they expect an object to fall to earth in a straight line when you let go of it? This is the question that scientists from Exeter, England took up (Fig. 8.6).

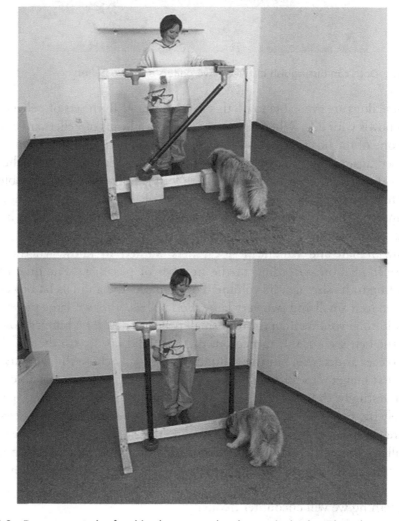

Fig. 8.6 Dogs expect the food in the cup under the vertical tube. They do not understand that the food is deflected by the diagonal tube

The experiment went as follows. There was an opaque tube in front of the test dog. It stood vertically and emptied into a food bowl. Next to it were two more food bowls, but they had no connection to the tube. Now a piece of food was thrown into the tube in plain sight of the animal. The dog could neither hear nor see where the piece came out. But it was allowed to look for it. The dogs decided correctly. That is, they always ran to the food container that was directly under the tube. Obviously, they expected the food to fall in a straight line.

But what happens when the tube is no longer vertical but diagonal? Then the food is deflected from its straight path. The reward then falls into one of the containers set to the side. The scientists confronted the dogs with this new situation. But they kept looking exactly under the spot where the food had been thrown in. They obviously did not understand that the food was deflected by the tube. After a few test runs, they could learn to go to the correct container, but only if it was always in the same place. Then they probably went back to the simple spatial rule: 'Always go right' (or 'Always go left!'). Because if the tube alternated, sometimes to the left and sometimes to the right, the dogs again made the wrong choice. They did not understand the mechanism. That is, they did not understand *how* a diagonally running tube affects the laws of gravity.

Once again, however, the dogs did find a strategy to solve the problem almost every time. They expect things to fall straight down. This is also what normally happens. When it doesn't, they look for a spatial rule.

Pulling Rope

Another experiment, also at Exeter, was conducted to test whether dogs understand certain relationships (Fig. 8.7). A piece of food was out of reach. It was in a box covered with mesh. So the dogs could see and smell it, but they

Fig. 8.7 Dogs do not follow the rope back to the food. They pull on the end of the rope that appears closest to the food

could not reach it directly. The food was attached to a cord. The other end of the cord extended beyond the box and was connected to a piece of wood.

The first part of the study tested whether dogs learn to pull the food towards them by using the piece of wood. All the test animals learnt to do this without much difficulty. They sometimes pulled the piece of wood with their snout, but mostly they used their paws. Over the course of the ten runs, they became more skilled at this and became faster and faster at solving the problem.

In the second part of the study, the task became more difficult. Now it was no longer a matter of dexterity alone, but also of understanding the situation. Now there were the ends of two cords lying there. But only one of them was connected to food. The ropes were either parallel or crossed. To get the reward, the dogs had to trace the cord to its end. Of course, they were supposed to choose the one with the food on it. But they didn't. They simply pulled on the end of the cord that was closest to the food. Most likely, there is a strong incentive to get their paws and snout as close as possible to the food. This strategy often leads to the goal—but not if the two cords are crossed!

From this experiment, the scientists concluded that dogs had not grasped the principle that the piece of food was connected to the rope. They did not trace the rope back to the food in order to choose the right end. However, there was a flaw in this experimental design. In most of these trials, a wrong choice means no reward for the dogs. Here, however, they were allowed to pull on the ropes until they got to the piece of food. This means that if they initially made the wrong decision, they could still pull on the other end and ultimately get to the food. In this way, there was not as much incentive to immediately pull on the right end. Maybe some dogs would make a more flexible decision after a failure? Or maybe a dog would be so frustrated that it would no longer pull on the end of any rope! But you can also try this out at home.

Despite this inconsistency in the experimental design, the results are believable. And again it looks like this: The dogs can solve the problem most of the time. But they have not grasped its underlying principle.

Dog Logic

Scientists in Leipzig also asked themselves whether dogs can recognise certain relationships. This time it was about a noise—and the connection between cause and effect. Let's say you have a piggy bank. You can't look inside unless, of course, you break it. What do you do to see if there's money in it? You shake it. And the rattle of the coins tells you that there is something inside. By virtue of the sound, you deduce the existence of the money. Something just like this was what the dogs were supposed to do in the experiment. It was an object choice trial with two cups. In one of the two cups, a piece of dry food was hidden. But the dogs tested did not know which one. As described in detail in Chap. 6, the animals now receive a cue as to where the food is hidden. This time, however, it was not a communicative cue. In other words, the person involved did not behave as if he or she wanted to consciously communicate something to the dog. Instead, it was a causal cue. One could understand it if one recognised the connection between cause and effect. Like the piggy bank, the cup with the piece of food was shaken. A sound was produced. At first it seemed as if the dogs could solve the problem: they chose the correct cup.

There was, however, a second situation. Now the empty cup was shaken. Thus, the dogs needed to understand that *no* food could be in this cup because *no* sound was made. Furthermore, they then had to conclude that the reward must be hidden in the other cup. Great apes can solve this task. The dogs, however, failed. They even preferred the empty cup. Probably because the human had touched it and shaken it. It is likely that they did not understand the principle that the shaken food produces the sound. Instead, they looked for a social cue. And then ran to the cup that was interesting because the person had touched it. Wolves, on the other hand, perform better at this task: they obviously understand the causal relationship—that the food in the shaken cup produces the sound.

Dogs are only able to solve this problem in an easier version (Fig. 8.8). Once again, a piece of food or a toy is hidden in one of two cups. Now it is shown that one cup is empty. At this point, the dogs can draw the logical

Fig. 8.8 The Border collie draws a conclusion: if one cup is empty, the reward must be in the other cup

conclusion: the reward must be in the other cup. But here too, the dogs only solve the problem if they are not influenced by the human involved. If the latter gives emphasis to one of the two cups by touching it, shaking it, or looking at it, then the dogs rely on the human cue rather than logic. These results indicate once again that dogs might not do so badly on some tests if they didn't allow themselves to be influenced by humans!

Do Dogs Know What They Smell?

As we have seen, dogs often don't do so well in tests about the physical environment. But most of the studies we have looked at so far are based on a sense that is highly important to us humans, and that is the sense of sight. Yet we actually know that dogs perceive their environment mainly through their nose, i.e., through their sense of smell (Fig. 8.9). One could therefore conjecture that they understand quite a bit about their environment as soon as smell plays a role. However, this is extremely difficult to test because we humans cannot smell well. We process smell rather unconsciously.

One thing we do know more about is that different smells play an important role when dogs communicate with their fellow dogs. In addition, they use their nose to distinguish edible from inedible food. It is commonly said that dogs can smell about 10,000–100,000 times better than humans. This means that they can detect unimaginably low concentrations of chemicals. They can also easily learn to recognise different smells. Specially trained dogs signal the presence of specific substances. For example, they are capable of recognising odours as diverse as those from drugs, explosives, plants, parasites, and corpses, or of detecting various diseases such as cancer and diabetes (see

Fig. 8.9 For dogs the sense of smell is very important

also Chap. 2). Finally, dogs are able to compare several odours with one another. They can tell whether two smells come from the same source or not.

Dogs are often used in real life by hunters or police officers to follow scent trails. This raises the question of how dogs use their sense of smell to find a target. A study by Irish scientists from Belfast investigated how dogs determine the direction of a scent trail left by a person. The researchers used a T-shaped test setup for this. A person walked from A to B, following the top line of the 'T'. After an hour, trained tracking dogs were let loose at the base of the 'T'. They were to find out whether the person had walked from A to B or from B to A. It is assumed that dogs succeed in doing this by searching for the direction in which the concentration of odour particles increases or decreases. That is, they turn in the direction in which the odour becomes stronger. The authors of the study wanted to know how many human steps the dogs need to determine the direction without error. The answer is: five. It only takes five human steps for trained tracking dogs to identify beyond doubt in which direction a person has walked. That is truly amazing!

Thus, we at least know something about what dogs can smell. But we have almost no idea about how cognition and smell are related to each other. In Jena, one of the first studies was conducted to understand how dogs perceive and understand the world through their noses. The scientists wanted to know whether dogs have an idea of what they are smelling. When dogs smell something, do they have an image in mind?

For this comparatively elaborate study, nearly 50 toy-loving dogs were tested, half of which were regularly deployed by the police or as part of a rescue dog team (Fig. 8.10). The study began with a pre-test in a small room. There, two toys were chosen for each dog that it liked, preferably to the same degree, so perhaps a ball and a ring. Then the experimenter left the small room with both toys, letting one toy disappear into a box. With the other, he or she made a trail in the test rooms by rolling it along the floor. Finally, the experimenter hid it behind one of four small barriers. Now the dog entered the room with its owner. The owner, who did not know which toy had been hidden, showed the dog the beginning of the trail and sent it off with the words 'Fetch it!'

There were two possible test conditions in the experiment. Either the scent trail came from the toy that was in the hiding place. That was the normal condition. Or at the end of the trail lay the toy from which the smell did *not* come. This was the surprise condition. The scientists wanted to know whether the dogs would show surprise or not. And the tested animals did indeed behave differently under the surprise condition than under the normal one. Especially in the first run of the surprise condition, they did not fetch the toy

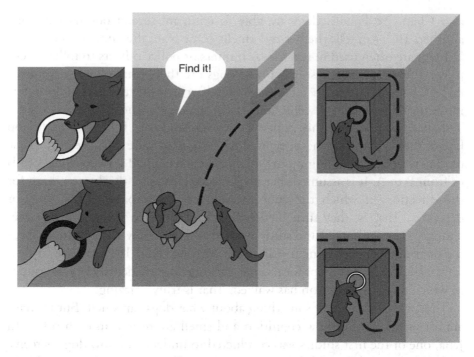

Fig. 8.10 Dogs have an expectation of what is at the end of a trail. They are surprised when the toy at the end of the trail is not exactly the same one that produced the trail

immediately. Rather, they continued to search. The scientists suspected that the dogs were looking for the right toy, namely the one from which the trace came. Thus, they were able to tell when the trace did not match the hidden toy. This suggests that during the test the dogs actually had a precise expectation of what the scent trail should lead them to. Thus, when dogs pick up a scent, they obviously do not only distinguish between good and bad or interesting and boring scents. Instead, they presumably really do develop some kind of idea of what awaits them at the end of the trail.

The Jena scientists took a closer look at the dogs' search behaviour. Unsurprisingly, animals with training retrieved the toys more quickly. Family dogs without training, which initially performed worse, improved their performance so much in the course of the tests that by the fourth test run they were just as fast. This once again shows their great learning ability.

The Jena researchers were also able to identify certain strategies. For example, the dogs were more likely to find the toys behind the two middle barriers by using their sense of sight rather than their sense of smell. For toys behind the right and left barriers, on the other hand, they consistently used their sense of smell. When and why dogs decide to use one sense or the other in a search remains to be investigated.

Vital for Survival!

Dogs show remarkable abilities when it comes to sniffing something out. In most of the other studies described in this chapter, however, the dogs used specific strategies to solve problems. Often, however, they did not *understand* the interrelated processes. This leads us to the question, why do they perform relatively poorly in many such tests about their inanimate environment?

There are three different possibilities. We do not yet know which is true. Perhaps canines just can't do it. Perhaps their closest relative, the wolf, would also fail such tests. We do not know how most animal species would perform in these tests. So far, it is mainly great apes that have been tested. They solve most of these tasks and seem to understand the relevant relationships. So it is possible that these abilities are actually the exclusive domain of humans and their closest relatives.

The second possibility would be that wolves can solve these problems. That is why it would be so interesting to test the dog's wild cousin. According to this theory, dogs have lost these abilities in the course of domestication. New studies suggest that this might be the case. For example, dogs do not understand that food in a cup makes a sound when shaken. Wolves do.

There is yet a third possibility. Wolves can, and so can puppies. But in the course of their lives with humans, dogs unlearn these connections. Because they don't *need* them. And that brings us to an especially important point.

For the question of why they do poorly is not really the most important one. It should be less about whether dogs understand certain principles. It should be about how they can survive *without* understanding those principles. Let's summarise again what the studies described have shown: Dogs understand that the ball continues to exist, even if they cannot see or smell it. But why do they have to imagine it being moved in a container? It is actually a very sensible strategy to always search for something in the place where it has disappeared from. And if there is nothing there, they then go over to where it was the last time. If all that doesn't help, their sensitive nose can help in the search.

Dogs are particularly good at orientating themselves in space. That was certainly important for their relative, the wolf, and it obviously also remains important for the domestic animal, the dog. A basic understanding of numbers could be useful: A person has left two steaks on the kitchen table, and then taken two away again. As a result, there are none left there. Regrettably. It is also certainly helpful if a dog knows that an apple falls vertically from a tree. Then it can get out of its way. But diagonal tubes that change the path of a falling object are rare. It is also a good tactic to scratch with one's paws as close as possible to inaccessible food. And if you don't know the causal relationship between sound and food, it is a good strategy to head for the shaken cup.

Dogs do not understand many of the processes described because they do not *need* to (Fig. 8.11). Most of us also have no idea how a computer really works, and yet we can operate one. With the help of different strategies, dogs also reach their goals. And thanks to the help of humans. Because dogs rarely have to solve causal problems themselves in our environment; we humans do that for them. They don't need to hunt for their food themselves, nor do they have to dig the litter den themselves. Why should they waste energy on something like that? Things work fine without it!

The most important thing for them is the human being. And that is why their extraordinary abilities lie more in the social domain. Observing people and making use of their cues, that is where our domestic dogs' strength lies. In fact, they are sometimes so fixated on people that it distracts them from solving a problem on their own.

You can even go a step further. One can imagine that it may not make sense at all for a dog to grasp all the processes in the human environment. Because it is often quite impossible to recognise logical connections. And that brings us back to Benny, the sheepdog, who is not surprised when his owner is no longer behind the door. One explanation for his behaviour would be, as

Fig. 8.11 Dogs can survive very well without understanding certain processes in their human environment

mentioned, that he understands exactly how a lift works. That he knows how the lift works with the help of an electric motor. We can't rule it out. But it is not likely.

However, we do know that Benny has ridden a lift many times. Perhaps he really was 'bewildered' the first time that the lift door opened again and the environment had suddenly changed. But if he tried to understand everything that was happening around him, it would probably just be confusing or even frightening for him. Like many other dogs, Benny is not only familiar with lifts, but also with cars and trains. Voices come from the radio, even though no human is there. Objects appear and disappear from the flat while he is away. The same door can sometimes be unlatched by him and sometimes not. Not 'wondering' about all of this can also have its benefits.

9

When Dogs Help

Almost everyone knows the story of the female collie, Lassie, who helps people when they are in danger. Lassie rescues people from the water, she alerts them to acute dangers, she fetches help when someone is injured. And maybe she would even share her food with humans. Lassie helps people and does so completely voluntarily, without expecting a treat.

Is Lassie just pure fiction? Or are dogs really that selfless?

To clarify this, we first have to say exactly what we mean by 'helping'. Let's put it this way: helping means that an actor (in our case Lassie) actively assists a recipient (in our case various film heroes) in some way. To do so, the actor invests energy in no small measure.

Biologists ask themselves about the evolution of such behaviour. How can it be that someone invests his valuable energy for another on his own accord? Darwin's theory of evolution can be reduced to a simple formula: the strongest, healthiest, and fittest individuals survive. Only these can pass on their genes. But in the story, one individual—namely the dog Lassie—voluntarily gives up some of her fitness. Why could such behaviour develop? The answer seems as logical as it is simple: if at the end of the day the helper does receive a benefit, even if it is only indirect, then his efforts for others are worthwhile. So-called kin selection is a classic example of the indirect benefit of helping. Individuals support relatives because in doing so they also ensure that their own genes are disseminated. This mechanism works, for example, in a wolf pack. There, older siblings as well as uncles and aunts participate in the rearing of pups.

But in our initial example, the film dog Lassie is not related to the injured at all. Lassie therefore helps more according to the motto: You help me, I help

J. Bräuer, J. Kaminski, *What Dogs Know*, https://doi.org/10.1007/978-3-030-89533-4_9

you. In biology, this is called reciprocity. Lassie invests energy now because she can assume—and rightly so—that people will return the favour later. But she probably doesn't consciously decide to help so that people will help her later. Instead, the motivation to help is innate, as it is with us humans. We too do not always act consciously and calculate how we can increase our fitness by helping others. Rather, mutual aid as a form of social behaviour has been beneficial to humanity as a whole and has therefore come to prevail in evolution.

An interesting question is also the personal reason for assisting others. Does Lassie help on purpose? Does she understand what she is doing? Helping someone presupposes two things on your part: understanding and motivation. The task that the helper needs to accomplish consists of understanding social situations. First of all, he or she has to recognise the goal of the other. That means, for example, that Lassie would have to understand that a swimmer aims to reach a safe shore. Secondly, Lassie would have to understand how this goal could be reached with as little exertion as possible. Lassie would therefore have to come up with the idea of pulling the drowning swimmer to the nearest shore.

In the chapter that follows, we will look at different types of helping: providing information, giving practical help, sharing fairly, and helping each other. In particular, we want to ask ourselves whether dogs that help understand human intentions and whether they are intrinsically motivated. For it is always possible that dogs have simply been trained to follow certain commands or to respond to certain situations in a certain way. To come back to our film dog: we want to answer the question whether Lassie is simply a well-trained dog or a truly authentic specimen of her breed.

Providing Information

Helping is not just about pulling someone out of the water or sharing food with them. It can also be highly beneficial for the recipient to receive vital information.

We now know that dogs are particularly good at making use of human cues (see Chap. 6). But does this also work the other way round? Can dogs also provide information to humans? A number of studies have shown that dogs inform unknowing humans about the location of hidden food or toys without having received any special training (Fig. 9.1). However, in all these studies, dogs only informed people about items that are important to themselves. Their behaviour is therefore beneficial to themselves and not to humans.

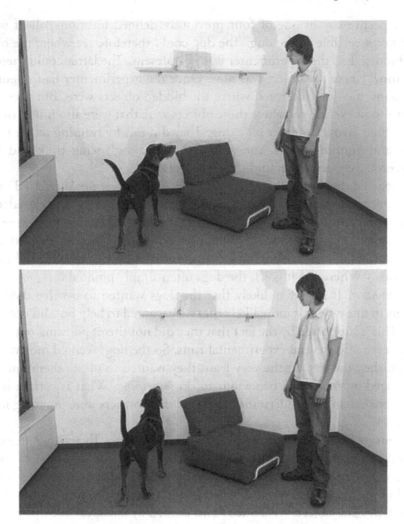

Fig. 9.1 Even untrained dogs inform their owners where a toy is hidden, e.g., by glancing back and forth between the toy and the person

However, we defined helping differently. Scientists from Leipzig therefore investigated whether dogs would help an unknowing individual to find a hidden object. The researchers varied whether the respective object was valuable for dogs, for humans, for both, or for neither. Depending on the condition, they chose a favourite toy (interesting for dogs), a play ring (interesting for both), a hole punch (interesting for humans), or a vase (interesting for neither). At the beginning of the experiment, the experimenter and/or the dog were occupied with the respective object. Then the experimenter left the room because he or she was distracted. A second person entered the room and hid the respective toy in one of four previously defined locations, all of which were not accessible to the dog. The dog could therefore see where the object was hidden, but the experimenter was not present. The latter could therefore not know where the object was now. Once the experimenter had re-entered the room, the dogs indicated where the hidden objects were. But they indicated almost exclusively where those objects were that were also interesting for them: their favourite toy or play ring. They did this by running to the hiding place and jumping up on the wall or by always glancing back and forth between the experimenter and the hiding place.

Rarely did the dogs point out those objects they were not interested in: the hole punch or the vase. It did not matter whether the person really needed the object or not. Especially when their own owner was the one they were supposed to inform about the hiding place, they sometimes behaved a bit strangely. While the owners looked around frantically and loudly addressed their dog ('Where did it go?'), the dogs often simply pointed out *some hiding place or other*. It is very unlikely that the dogs wanted to deceive their own owners in this way. It is more likely that they wanted to help but did not know *how*. This is supported by the fact that they did not tire of pointing out hiding places even after several experimental runs. So the dogs seemed motivated to inform their owners. At the very least, they wanted to please them and were stimulated by the search behaviour to do *something*. What is certain is that they were not able to understand what goal their owners were pursuing in this situation.

Scientists from Portsmouth have taken up the issue. They chose an experimental design in which three objects played a role. The first object, a notebook, was of interest to the person. The second object, a dog toy, was naturally exciting for the dog. The third object was picked randomly, in this case, an office item. The objects were once again hidden by a helper in the presence of the dog. Again, the experimenter did not know where the objects were. This time, however, there were only two hiding places. And the helper sometimes used both hiding places at the same time. In all three test conditions, he or she first hid the dog toy and then either the notebook, the random object, or nothing at all. Again, the experimenter entered the room only after the helper had hidden the objects. He or she held a pen in his/her hand because he/she was looking for his/her notebook. It turned out that the dogs most often pointed to their toys. However, they did distinguish between the notebook and the other objects. They pointed out the notebook to the experimenter much more persistently.

In another experiment, no toy at all was hidden, but instead either the notebook or the random item. Here, too, the dogs pointed to the notebook for longer. They especially did so when the experimenter spoke to them in a high-pitched voice to express his despair about the missing notebook (see Chap. 7). From this we can see that the dogs were again very motivated to point out an object that was hardly interesting in itself. This time they even distinguished whether the object was important for the person or not. But exactly how they understood the situation and whether they really wanted to consciously provide helpful information is not yet clear from this.

Interestingly, in a similar test, great apes exhibited completely different behaviour. They also pointed most often to objects that they could use themselves. In the first runs of a behavioural study, they even pointed to the object that was only interesting for the experimenter. But as soon as they realised that there was nothing more in it for them, they simply stopped doing this. From this we can conclude that great apes are quite capable of understanding situations and imagining what humans are looking for. But they are hardly motivated to help.

In contrast to great apes and dogs, we humans are usually highly reliable in providing our fellow humans with a wide variety of information. Small children already start doing this at the age of 1 to 2 years. In similar experiments, they pointed out various objects that adults needed in certain situations. The children did this regardless of whether they benefited from their behaviour or not. It also turned out that their motivation did not diminish over many test runs. So we can be sure that young human children have both: an understanding of situations in which others find themselves and the motivation to provide helpful information.

Practical Help

Now let's talk about practical help. The question is whether actors help their respective recipients in such a way that it actually benefits them. So when film dog Lassie pulls a person out of the water, would other dogs act similarly? Or does something like that only happen in the movies?

Researchers in Leipzig wanted to know whether dogs would help open a door (Fig. 9.2). As we know, dogs sometimes have problems understanding the aims and intentions of others. The idea was therefore to create a situation that was as obvious as possible to the dog and as clear as possible in terms of intention. The setup was as follows: An individual tried to enter a locked room in order to pick up his key, which was lying on the floor. This room was surrounded by waist-high plexiglass walls and had a transparent plexiglass door. The dogs to be tested learnt to open the door by pressing down on a

Fig. 9.2 Dogs open a door to a room when they realise from the person's behaviour that he or she wants to get into the room

floor switch before the experiment began. During the experiment, there were three sets of test conditions: the help condition, the control condition, and the food condition. In the help condition, the person expressed physically that he or she wanted to enter the locked room. To do this, the person shook the door, bent over the plexiglass wall to reach for the key, and spoke to the dog, telling it aloud about his/her problem. Of course, the person never used a command known to the dog or pointed to the switch. The help condition was compared with the control condition. In this one, the key was again out of reach in the locked room, but the person was simply reading a book and not interested in it. Obviously, he or she did not need the key. Finally, in the food condition, the person was again reading a book. This time, however, instead of a key, there was delicious food in the locked room. Under this condition, the dogs almost always opened the door independently and without being asked. The dogs opened the door just as often to help the person. In the control condition, on the other hand, they rarely opened the door. So if neither human nor dog had a reason to enter the locked room, the latter did not do so.

Now, it could be assumed that the dogs in the help condition simply exhibited behaviour they had previously learnt. They opened the door because they knew how to do it and because they noticed the person was agitated. This assumption, however, turned out to be rather unlikely after further experiments with agitated persons. Moreover, the dogs opened the door over several test runs without receiving a reward for doing so. Therefore, the scientists felt confirmed in their assessment that dogs are basically very motivated to help, but that they often have problems recognising human intentions if these are not clearly communicated.

However, dogs' problems are not limited to their failure to recognise human intentions. In some situations, they may also lack the knowledge of how they can even help. Researchers from Ontario in Canada wanted to know whether dogs seek help from others when their owners are in distress. In the study, the dog owners feigned a heart attack. A witness observed the situation from close by. The dogs had the opportunity to get to this witness and ask him or her to help. Unlike our film dog Lassie, the dogs in the study did not seek the help of the witness. It is likely that the dogs did not really know how to help in this situation. It is also possible that they did not see the situation as an emergency. The owners were just pretending to have a heart attack. The emergency was not real, which the dogs presumably realised.

There is growing evidence that dogs do react to real epileptic seizures or sugar shocks in diabetics (see Chap. 2). They may even be able to predict them, even if they have not been trained to do so. It is not entirely clear whether dogs are motivated to help their owners in such situations or whether they are merely agitated because their owner smells and behaves unusually. In any case, in the study with the mock heart attack, the authors reported that the dogs watched their owners closely as they feigned the emergency. This at least suggests that the dogs were concerned about their owners. On the other hand, it could just be that the dogs found their owners' behaviour strange or frightening.

A comparison with chimpanzees shows that they also help fellow members of their species. They open the door for an unrelated member of their group or provide them with a tool they need. They can even guess which tool the recipient needs to get to the food it seeks. Hand-raised chimpanzees even help their human adoptive parents when the latter attempt to reach for something out of their reach. This means chimpanzees are able to discern the aims of potential recipients of help and are also motivated to provide it. Even if dogs and apes do help in certain circumstances, there can be no doubt that we humans are unique in the way we help each other. Toddlers as young as 2 years old willingly help other children and unrelated adults in a variety of situations. It seems to be an inner need of theirs.

Sharing Is Fun?

Sharing food could be seen as another way of helping each other. We know that adult animals regurgitate already chewed food to feed their offspring. Apart from this behaviour, active sharing between animals is not usually observed. However, it is more common for animals that have food in their possession to be tolerant enough to let other members of their species eat with them. As we saw in Chap. 3, dogs are worse at this than wolves.

A closely related question is whether and how dogs provide food for others when they do not get anything for themselves in return. In comparative psychology, we speak of pro-social behaviour in such cases. The idea is this: Imagine you have finished reading a magazine and now have the choice of either throwing it away or giving it to someone else. How do you decide? In most cases, you would certainly pass on a magazine you have read. After all, it costs you nothing and earns you social prestige points. The friend or colleague to whom you have given the magazine will certainly find occasion to reciprocate. But how do dogs act in similar situations?

Scientists test such behaviour by having individuals choose between selfish and pro-social options. Logically, this requires two test participants, one of whom must decide (the test dog) and another who is at the mercy of the decision (the recipient). In the pro-social variant, both leave the situation with a reward; in the egoistic variant, the recipient who is dependent on the other's decision gets nothing. In the actual experiment, food was presented on a tray. The test dog could pull the tray towards it and thus provide itself and the recipient with food. The test dog had a choice of two trays, one containing food for itself and the recipient, and the other containing food only for itself. It turned out that the dogs preferred the pro-social variant. They chose the tray that contained food for both dogs. The test dogs made their behaviour dependent on their relationship to their fellow dogs. They only 'gave away' food if they knew the other dog. They did not do this for an unknown dog.

It could be argued that dogs see humans as their actual social partners. Scientists in Vienna therefore carried out the test just described in such a way that people now took on the dependent role of the recipient. Surprisingly, the dogs did not show any pro-social behaviour whatsoever in this case. They also did not distinguish between recipients they knew and those they did not. However, it was noticeable that the test dogs looked at the dependent humans significantly longer than at dependent members of their own species. This could be explained, for example, by the fact that the animals were waiting for a signal. Perhaps they found it unusual that the humans did not communicate with them at all. More generally, these results suggest that dogs either have problems understanding such demanding situations or lack the motivation to give away food. Chimpanzees, by the way, also do not provide food to their fellow chimpanzees in similar situations, even if it costs them nothing.

An interesting question at this point is whether animals have a sense of fairness. Do animals find it unfair when another animal gets more or better food? Scientists, in an effort to get closer to an answer, have tested monkeys. Their experiment has attracted a lot of media attention. In short video sequences that can be viewed on YouTube, capuchin monkeys are visibly annoyed that fellow members of their species receive better food. Unfortunately, these studies are methodologically flawed. As a result, experts tend to doubt that animals have a sense of fairness comparable to ours.

To explore this question, a group of scientists in Vienna tested dogs. The dogs had to perform a task that consisted of giving a paw or pressing a button. They were rewarded for this with food that was acceptable but not really tasty. When they had finished, a fellow dog within sight was given either the same food or much better food for completing the same task. The result was that the test dogs completed their tasks without any further interest in what food they or their fellow dogs got. Wolves and enclosure dogs, however, seemed to draw a distinction, but the effect was very slight. What all animals did poorly with, however, was when they were not rewarded at all. This seems, at first glance, logical enough. After all, people's morale decreases when we are not rewarded at all for our efforts. But that has nothing to do with fairness. It was much more interesting that the dogs distinguished between situations in which the partner was rewarded and situations in which there was no partner at all. In situations where there was no reward, their enthusiasm diminished much more quickly if a fellow dog was present who was rewarded. Whether or not one can therefore speak of a sense of fairness is still a matter of debate among experts. The presence of a fellow member of the same species seems to at least play a role. 'Working morale' decreases more quickly when the fellow dog is rewarded. It could also be that the dogs stop participating because they expect food themselves when they see others being rewarded.

But it is not only food that can be shared, but also attention. There have been several studies that have caused quite a public stir, allegedly demonstrating jealousy in dogs. In these studies, a fake-dog was used that received food or affection from a person. The scientists then observed whether dogs exhibited a typical jealous response. For example, did they try to attract the human's attention or attack rivals. However, it is completely unclear whether the tested dogs even perceived this artificial rival as a real dog. If you sometimes experience in your daily life that your dog harasses you while you are petting your neighbour's animal, then there is a much simpler explanation. Presumably, in such a situation, you give—through your body language and speech—the signal that normally indicates that it is now time for physical contact. You can safely assume that your dog will respond to this.

It cannot be completely ruled out that dogs do in fact feel jealousy. It just hasn't been proven yet. Scientists in Leipzig have carried out a study whose results speak against this. The study looked into the question of how dogs judge people. Our companion can, of course, distinguish whether someone is friendly to it or not. Dogs remember, for example, who has played ball with them once. When they meet this person again, they try to animate them to play. But what happens when test dogs see how a dog from their household is treated? The experimental design is classic. There is a 'loving' experimenter and a 'bad' one. The former pets the dog and plays with it, while the latter simply ignores it. How does a test dog behave who has observed this? Does it learn from the observation? No, it doesn't. It turned out that the dogs did not prefer the 'loving' experimenter. They obviously were only able to learn from their own direct experience, but not by observing what had happened to another dog. If this is so, then dogs lack the basic prerequisites for feeling fairness and jealousy in the way we do as humans.

Helping Each Other

We have now looked at different forms of help in which individual actors support specific recipients. Now we want to look at classical forms of cooperation. The question is whether and how dogs can pursue a common objective either with each other or with humans. In this case, both are actors and recipients in equal measure.

Jena scientists investigated whether and how cooperation works among dogs and wolves, respectively (Fig. 9.3). They designed a behavioural test that sought to recreate a hunting situation such as occurs when two predators jointly attempt to take down a larger herbivore, such as an elk. The approach assumed that in the wild, one of the predators must attract the attention—and the dangerous antlers—of the herbivore so that the other predator can attack and bring down the prey. In such a situation, it is imperative that the predator taking the greater risk in the hunt is confident that it will receive a fair share of the jointly hunted prey. The experimental setup consisted of a T-shaped barrier separating the animals from two filled food bowls. To the left and right of the barrier were closable openings. As soon as an animal approached one of the openings during a test run, it was closed while the

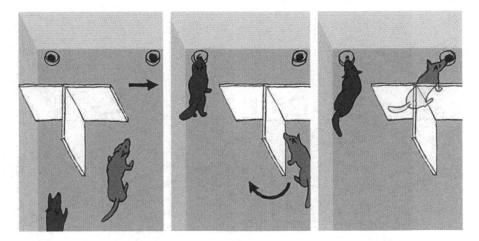

Fig. 9.3 In this task, the animals must coordinate their actions so that they are each on one side of the sliding door at the same time. As soon as one animal gets through the opening, the sliding door stops and the other animal can also get to the food through the same opening

other opening remained open. So the animals had to coordinate with each other so that they positioned themselves at the openings at the same time. The animal that had waited patiently could get to the food first, while the more active animal had to first circle the trunk of the 'T' to likewise be able to eat.

Dogs and wolves were equally successful at mastering this task. On average, they reached the food in three out of four cases. However, just like the wolves, the dogs did not always share the food. The more active animal, which still had to circle around the barrier, was sometimes at a disadvantage because it got to the food later. Pairs where the more dominant animal did the active legwork had an advantage. The more active animal potentially took the risk of not getting its share of the food. However, its higher rank ensured that it received its share even if it came to the food later than its lower-ranking hunting partner. This study illustrates how cooperation and competition are sometimes connected. The intention was actually to study cooperation. However, it turned out that competition for the reward also influenced the behaviour of the pairs.

The task that dogs and wolves had to solve in this study was relatively straightforward. The animals had to coordinate with each other to get to the food. They did not need training for this. However, wolves seem to perform better than dogs in more complex cooperative tasks. Scientists from Ernstbrunn near Vienna compared wolves and dogs at a task in which two dogs—or two wolves—had to pull a board together (Fig. 9.4). There were two rewards on

Fig. 9.4 To solve this problem together and get to the food on the board, the two cooperating partners have to pull on both ends of the rope at the same time

the board and, as the test begins, the board is out of reach. It was, however, connected to the animal enclosure by two loose rope ends in such a way that one animal could not reach both ends.

Both animals had to pull on both ends of the rope at the same time to reach the tray and thus the food. The Ernstbrunn wolves did much better at this behavioural test than the dogs raised in the enclosure. Family dogs were able to solve the task quite well. Again, it probably matters whether and how the food on the tray was shared. This is easier for wolves.

Beyond that, the task does not appear to be particularly easy for either dogs or wolves; it requires intensive training. As we know from Chap. 8, dogs are not generally good at deliberately pulling something towards them.

In fact, all the animals in the Ernstbrunn study had already gained extensive experience in pulling food on a rope towards them. It could therefore be that dogs perform comparatively poorly in this cooperative task because they only understand their physical world to a limited extent. They are also not as tenacious or persistent as wolves are.

We have already seen on many occasions that dogs' natural social partners are not their fellow dogs, but humans. For this reason, the scientists in Ernstbrunn also tested dog–person pairings and wolf–person pairings at the tray experiment. Dogs performed just as well as hand-reared wolves. Both had little trouble cooperating with people to bring the tray over. However, there were also remarkable differences between the species. The researchers summed up one of them as follows: wolves lead, dogs follow. The wolves were more inclined to call for cooperation and also to take the lead. Dogs, on the other hand, often waited for the person to act. They even behaved this way when they were closer to the board. They looked at the person to see what he or she was up to. In contrast to the wolves, the dogs also did not try to take one end of the rope away from their human partner. From this behaviour, the scientists concluded that dogs work so well with humans because they know how to judge them better than they can judge a fellow dog, and they willingly submit to them.

Finally, let's take another look at our famous film dog Lassie. Are dogs also smart and selfless in real life, or is this something that only exists in the movies? How much truth is there in the story of the tireless collie who selflessly helps people, always sees the big picture, and always knows what to do? To keep it short and sweet: It is not so much that our everyday dogs lack selflessness, because dogs are highly motivated to cooperate with us in a variety of situations. What is more likely to be lacking is 'smarts'. It is not uncommon for dogs to have problems understanding the situation in its entirety. They often simply do not know what should be done from a human point of view in order to really help. However, when humans give their dogs clear and unambiguous cues, dogs are very reliable helpers, even without food rewards.

10

Looking Ahead

We have learnt a lot about our dogs in the preceding pages. Some of it may have been new to you, some of it not. For some of the things you have long suspected, you now have scientific proof. And we hope that you were sometimes surprised by the feats of the dogs and wolves we presented to you.

We started our book describing what the early days of dogs were all about. Maybe you had only known the story of how the primeval hunter brings home a wolf pup. Now you have seen that, according to a new theory, maybe the wolf also domesticated itself to a certain extent.

You have also learnt exactly what happened when wolves became dogs. You now know that your dog has lost some of its progenitor's abilities but has at the same time acquired some new ones. The next time you have to leave your dog alone because you have something urgent to do, it will be clear to you why it is so difficult for it to stay behind alone. The need to be close to people is something dogs are born with.

You should no longer be surprised if your dog always seems to know exactly when the best moment is to take food from the table. The suspicion you may have long held has been scientifically confirmed. Dogs have a good sense of what we humans can and cannot see. They are very sensitive to when we are alert and when we are not.

But you have also learnt something about your pet's limitations. You may have been surprised to learn that dogs rarely learn by watching others. True, they can profit from watching others solve a problem. But they do not copy it exactly. This is a topic that always makes dog owners sit up and take notice. Didn't Labrador Ben learn to unlatch the door from his mother? But maybe he just scratched in the same place with his paws as she did. Or maybe the right test has not yet been developed to detect this kind of learning? Until this test is found, you may take comfort in the fact that apes are also bad at 'aping' something. For this exact copying of an action seems to be something typically human.

There is another area in which dogs have a hard time. They understand surprisingly little about their inanimate environment. The interesting thing is that they often develop strategies for solving the problems at hand. For example, they pull on a rope that is connected to a piece of food. But if you look more closely, it turns out that the dogs have not really understood the process. They just pull on the end of the rope that is closest to the food. They do not trace the rope back to the food. These strategies usually help to solve the problem. But not always! So don't be surprised if your dog keeps going back to an empty food container. If you shake it to show that there is nothing in it, your pet will not understand. But don't be disappointed. Other dogs can't do it either!

Various strategies often help, but even more often humans help. Dogs are so well adapted to us that they can hardly survive without us. They are dependent on us. That's why there is one area where our dogs are real specialists. And that is when it comes to communicating with us humans. For example, your dog is very good at indicating to you where his favourite toy is hidden. You don't even have to train it for that.

Perhaps even more impressive are its abilities when it comes to understanding our communication. Be it our gestures or even our spoken commands, dogs are very good at making use of our signals. So next time you communicate with your dog, remember how receptive it can be to the smallest of cues. It has been scientifically proven that it is even more sensitive to these cues than our closest relatives, the great apes. An animal that can even interpret our winking as a sign needs not to be shouted at to get it to understand. Maybe we have been able to help you to look at your dog with different eyes.

Many Questions Remain Unanswered!

This book cannot claim to be complete. Our aim was to give you a glimpse into the current state of science. And this can change quickly. A lot of research is being done in this area. Scientists all over the world are now conducting studies on the subject of cognition in dogs. Because dogs are easy to test. They are usually very uncomplicated and friendly. You can motivate them easily. And you don't even have to go to the zoo to do it. Because they are everywhere! Maybe next month a study will be published that will once again overturn one of our ideas about our four-legged friends.

Because a lot of questions are still unanswered. For instance, we still know very little about how dogs understand and interpret smells. Yet the nose is perhaps their most important sensory organ. But we humans are only just beginning to grasp what dogs might actually understand in this area. The facial expressions of dogs also still need to be systematically studied. We now have new and objective methods to look at exactly when and how dogs move their facial muscles. Perhaps a detailed ethogram of dogs' facial expressions can result from this research.

Why Is This Important?

Why do we take up this topic? For one thing, of course, because as dog lovers we are interested in what is going on inside our protégé. But this is not the only reason. Understanding what a dog understands is also important in terms of human development. One of the big questions is how human cognition has developed in the course of evolution. What is uniquely human. Many human abilities have long been thought to be unique. But the more closely we have studied these abilities, the more often we have found parallels in some animal species. This is important. It gives us insights into ourselves. About what has changed in the course of our development and what abilities we share with other species.

The dog is one of the most suitable candidates for studying such abilities. For example, it was thought that only humans are capable of learning new concepts by means of indirect matching. If this is now also found in an animal, there cannot be much to its uniqueness. It has now been proven that the Border collie Rico was able to learn new terms in exactly this way. So this is not an ability that only humans have developed. Rico was not human and thus the supposition has been disproved.

We have also learnt how good dogs are at using our pointing gestures to find food. Accordingly, this cannot be an exclusively human form of communication. Not even when our closest relatives have great difficulty utilising our gestures. It becomes particularly interesting when dogs develop certain abilities that are not found in any other animal species. Abilities that, apart from dogs, are only found in humans. For example, there is some evidence that our domestic dogs are *better* at using pointing gestures than any other animal. This could be an indication that dogs have developed special skills while living together with humans. During this long time together, certain things were probably nurtured and others not. One can, for instance, well imagine that the dogs that were particularly attentive to humans enjoyed an advantage. They responded better and were thus favoured. Perhaps they were given a little more food than the less attentive dogs and were better able to reproduce. As a result, this behaviour took hold.

These special abilities of dogs may even give us information about our own development. For example, about what might have influenced natural selection in humans. We most likely encouraged the friendly, attentive dogs that made contact with us. Perhaps friendly individuals in particular have also prevailed in human development. Maybe this is how some special abilities were formed. For example, it is typical for us humans to be very good at cooperating with one another.

It is still pure speculation whether we can draw such conclusions from the abilities of dogs. But perhaps these investigations will actually help us to learn something about ourselves.

Animal Wellbeing

But there is another—from our point of view—very important reason why we should study the cognitive abilities of dogs. The wellbeing of animals. If we do not understand what is going on in a dog's head, we run the risk of systematically misjudging them. When working with dogs, there are two extremes. One is so-called behaviourism. This is the assumption that changes in a dog's behaviour can only be brought about through conditioning and training. The underlying assumption is that dogs are hardly capable of flexible thinking. The other extreme is so-called anthropomorphism. This means something like the 'humanisation' of dogs. The assumption underlying this kind of thinking is that all behaviour that dogs exhibit that is similar to humans also involves human-like understanding. Those who think this way will be convinced that dogs understand everything we humans say and feel, that dogs are unquestionably empathic, and so on. What is certain is that neither of these extreme ways of thinking does justice to our dogs. Both behaviourism and anthropomorphism stand in the way of a true understanding of dogs.

Let's look at empathy, for example. When we say empathy, we mean recognising, understanding, and empathising with the emotional world of another individual. Can dogs do this? Can dogs recognise, understand, and empathise when we humans are happy or sad? Science cannot yet answer this question. A Viennese study was able to show that dogs react to emotional noises from humans and members of their own species, sometimes with stress. But if dogs are indeed stressed by human emotions, then we might have to think about using them in animal-assisted therapy. In such therapies, dogs are sometimes confronted with very emotional situations. Shouldn't we ask whether a rather sensitive dog is not exposed to a stress level in this way that might be contrary to the animal's wellbeing?

To better answer these and many other questions, we need science. We need to understand how dogs experience and understand different situations before we place our four-legged friends in them.

How Smart Is My Dog?

Dogs are an integral part of our everyday human lives. It may therefore seem remarkable how little we still know about them. We therefore hope that in this book you have learnt how important it is to question dog behaviour. People often assume that the 'thought' behind a certain way of behaving is similar to their own. In this book you have learnt that this need not necessarily be the case. However, this is not supposed to mean that your dog is any less lovable if it cannot understand certain things.

Because even without such an understanding, our domestic dogs have succeeded brilliantly. They have adapted perfectly to living together with us humans. Millions of them live in our towns and villages. And a glance at any shop for everyday pet supplies shows that we let our shared life with dogs cost us a pretty penny. But they are also useful to us. A study from Finland, for example, has convincingly demonstrated how useful a dog can be for hunters. The assistance of a dog increases the hunters' probability of success by more than 50%. This has probably been the case since humans and dogs began cooperating with each other tens of thousands of years ago.

Today, dogs act as reliable guides for the blind, as thorough search rescuers, and as dedicated herding dogs. In many cases, however, they are simply our companions. And every dog owner knows that it is scarcely possible to put a price on living with a four-legged friend.

But how smart are dogs? Perhaps you would have liked a precise answer to this question at this point. How smart is it? Does it have an overall grade of 'A' or 'F' on its report card? If you wish, there might be an 'A' for using pointing gestures, and a 'C-' for social learning. But we don't want to judge. After all, who should decide which skill is the most important? If a dog were to write a book, it might say, 'The deciding factor is a good nose.' In fact, dogs smell certain substances about 100,000 times better than we do. In addition, they are very good at remembering the individual smell of a human being or a fellow dog in order to then match it up to a particular individual. We can't do that. Does that make us dumber than dogs? Certainly not. Because each of us has developed exactly the abilities we need.

We hope you have learnt some things about dogs in this book that you didn't know. We hope we have been able to give you a little insight into what dogs understand about the world they live in. We would be even happier if we have been able to dispel some of the prejudices about the wolf's descendants in this way. They are neither 'better human beings' who understand everything nor beasts ruled by reflexes. What they are instead: fantastic companions.

Further Reading

How Wolves Became Dogs

Archer, J. (1997). Why do people love their pets? *Evolution and Human Behavior, 18,* 237–259.

Bonanni, R., & Cafazzo, S. (2014). The social organization of a population of free-ranging dogs in a suburban area of Rome: A reassessment of the effects of domestication on dog behaviour. In J. Kaminski & S. Marshall-Pescini (Eds.), *The social dog: Behaviour and cognition* (pp. 65–104). Elsevier.

Coppinger, L., & Coppinger, R. (2002). *Dogs: A new understanding of canine origin, behavior and evolution.* University of Chicago Press.

Kaminski, J., & Marshall-Pescini, S. (2014). *The social dog: Behavior and cognition.* Elsevier.

Kaminski, J., Waller, B. M., Diogo, R., Hartstone-Rose, A., & Burrows, A. M. (2019). Evolution of facial muscle anatomy in dogs. *Proceedings of the National Academy of Sciences, 116*(29), 14677–14681.

Serpell, J. (1995). *The domestic dog: Its evolution, behaviour and interactions with people.* Cambridge University Press.

Thalmann, O., Shapiro, B., Cui, P., Schuenemann, V. J., Sawyer, S. K., Greenfield, D. L., & Wayne, R. K. (2013). Complete mitochondrial genomes of ancient canids suggest a European origin of domestic dogs. *Science, 342*(6160), 871–874.

Waller, B. M., Peirce, K., Caeiro, C. C., Scheider, L., Burrows, A. M., McCune, S., & Kaminski, J. (2013). Paedomorphic facial expressions give dogs a selective advantage. *PLoS One, 8*(12), e82686.

Dogs Are Not Wolves

Cafazzo, S., Marshall-Pescini, S., Lazzaroni, M., Virányi, Z., & Range, F. (2018). The effect of domestication on post-conflict management: Wolves reconcile while dogs avoid each other. *Royal Society Open Science, 5*(7), 171553.

Marshall-Pescini, S., Rao, A., Virányi, Z., et al. (2017). The role of domestication and experience in 'looking back' towards humans in an unsolvable task. *Scientific Reports, 7*, 46636.

Miklosi, A., Kubinyi, E., Gacsi, M., Viranyi, Z., & Csanyi, V. (2003). A simple reason for a big difference: Wolves do not look back at humans but dogs do. *Current Biology, 13*, 763–766.

Nagasawa, M., Mitsui, S., En, S., Ohtani, N., Ohta, M., Sakuma, Y., ... Kikusui, T. (2015). Oxytocin-gaze positive loop and the coevolution of human–dog bonds. *Science, 348*(6232), 333–336.

Parker, H. G., Kim, L. V., Sutter, N. B., Carlson, S., Lorentzen, T. D., Malek, T. B., et al. (2004). Genetic structure of the purebred domestic dog. *Science, 304*, 1160–1164.

Range, F., Brucks, D., & Virányi, Z. (2020). Dogs wait longer for better rewards than wolves in a delay of gratification task: But why? *Animal Cognition, 23*(3), 443–453.

Svartberg, K., Tapper, I., Temrin, H., Radesater, T., & Thorman, S. (2005). Consistency of personality traits in dogs. *Animal Behaviour, 69*, 283–291.

Topal, J., Gergely, G., Erdohegyi, A., Csibra, G., & Miklósi, A. (2009). Differential sensitivity to human communication in dogs, wolves, and human infants. *Science, 325*(5945), 1269–1272.

Zimen, E. (1990). Der Wolf: Verhalten, Ökologie und Mythos. Knesebeck & Schuler.

What Do Dogs Understand About Others?

Call, J., Bräuer, J., Kaminski, J., & Tomasello, M. (2003). Domestic dogs are sensitive to the attentional state of humans. *Journal of Comparative Psychology, 117*, 257–263.

Kaminski, J., Bräuer, J., Call, J., & Tomasello, M. (2009). Domestic dogs are sensitive to a human's perspective. *Behaviour, 146*(7), 979–998.

Kaminski, J., Pitsch, A., & Tomasello, M. (2013). Dogs steal in the dark. *Animal Cognition, 16*(3), 385–394.

Do Dogs Learn by Observing Others?

Bhadra, A., & Bhadra, A. (2014). Preference for meat is not innate in dogs. *Journal of Ethology, 32*(1), 15–22.

Kubinyi, E., Topal, J., Miklosi, A., & Csanyi, V. (2003). Dogs (*Canis familiaris*) learn from their owners via observation in a manipulation task. *Journal of Comparative Psychology, 117*, 156–165.

Mersmann, D., Tomasello, M., Call, J., Kaminski, J., & Taborsky, M. (2011). Simple mechanisms can explain social learning in domestic dogs (*Canis familiaris*). *Ethology, 117*(8), 675–690.

Pongracz, P., Miklosi, A., Kubinyi, E., Gurubi, K., Topal, J., & Csanyi, V. (2001). Social learning in dogs: The effect of a human demonstrator on the performance of dogs in a detour task. *Animal Behaviour, 62*, 1109–1117.

Tennie, C., et al. (2009). Dogs, *Canis familiaris*, fail to copy intransitive actions in third-party contextual imitation tasks. *Animal Behaviour, 77*(6), 1491–1499.

How Do Dogs Interpret Human Gestures?

Bálint, A., Faragó, T., Meike, Z., Lenkei, R., Miklósi, Á., & Pongrácz, P. (2015). 'Do not choose as I do!' – Dogs avoid the food that is indicated by another dog's gaze in a two-object choice task. *Applied Animal Behaviour Science, 170*, 44–53.

Bräuer, J., Kaminski, J., Riedel, J., Call, J., & Tomasello, M. (2006). Making inferences about the location of hidden food: Social dog, causal ape. *Journal of Comparative Psychology, 120*(1), 38–47.

D'Aniello, B., Scandurra, A., Alterisio, A., Valsecchi, P., & Prato-Previde, E. (2016). The importance of gestural communication: A study of human–dog communication using incongruent information. *Animal Cognition, 19*(6), 1231–1235.

Hare, B., Brown, M., Williamson, C., & Tomasello, M. (2002). The domestication of social cognition in dogs. *Science, 298*, 1634–1636.

Riedel, J., Schumann, K., Kaminski, J., Call, J., & Tomasello, M. (2008). The early ontogeny of human-dog communication. *Animal Behaviour, 75*(3), 1003–1014.

Communication Between Dogs and Humans

Ben-Aderet, T., Gallego-Abenza, M., Reby, D., & Mathevon, N. (2017). Dog-directed speech: Why do we use it and do dogs pay attention to it? *Proceedings of the Royal Society B: Biological Sciences, 284*(1846), 20162429.

Faragó, T., Pongrácz, P., Range, F., Virányi, Z., & Miklósi, Á. (2010). 'The bone is mine': Affective and referential aspects of dog growls. *Animal Behaviour, 79*(4), 917–925.

Kaminski, J., Call, J., & Fischer, J. (2004). Word learning in a domestic dog: Evidence for 'fast mapping'. *Science, 304*, 1682–1683.

Kaminski, J., Hynds, J., Morris, P., & Waller, B. M. (2017). Human attention affects facial expressions in domestic dogs. *Scientific Reports, 7*(1), 12914.

Mitchel, R. W. (2004). Controlling the dog, pretending to have a conversation or just being friendly. *Interaction Studies, 5*, 99–129.

Pongracz, P., Molnar, C., Miklosi, A., & Csanyi, V. (2005). Human listeners are able to classify dog (*Canis familiaris*) barks recorded in different situations. *Journal of Comparative Psychology, 119*, 136–144.

Yin, S. (2002). A new perspective on barking in dogs. *Journal of Comparative Psychology, 116*, 189–193.

What Do Dogs Know About Their Environment?

Bräuer, J., & Belger, J. (2018). A ball is not a Kong: Odor representation and search behavior in domestic dogs (*Canis familiaris*) of different education. *Journal of Comparative Psychology, 132*(2), 189–199.

Hepper, P. G., & Wells, D. L. (2005). How many footsteps do dogs need to determine the direction of an odour trail? *Chemical Senses, 30*(4), 291–298.

Lampe, M., Bräuer, J., Kaminski, J., & Virányi, Z. (2017). The effects of domestication and ontogeny on cognition in dogs and wolves. *Scientific Reports, 7*, 11690.

Miletto Petrazzini, M. E., & Wynne, C. D. (2016). What counts for dogs (*Canis lupus familiaris*) in a quantity discrimination task? *Behavioural Processes, 122*, 90–97.

Osthaus, B., Lea, S. E. G., & Slater, A. M. (2005). Dogs (*Canis lupus familiaris*) fail to show understanding of means–end connections in a string-pulling task. *Animal Cognition, 8*, 37–47.

Watson, J. S., Gergely, G., Csanyi, V., Topal, J., Gacsi, M., & Sarkozi, Z. (2001). Distinguishing logic from association in the solution of an invisible displacement task by children (*Homo sapiens*) and dogs (*Canis familiaris*): Using negation of disjunction. *Journal of Comparative Psychology, 115*, 219–226.

When Dogs Help

Bräuer, J. (2015). I do not understand but I care: The prosocial dog. *Interaction Studies, 16*(3), 341–360.

Bräuer, J., Schönefeld, K., & Call, J. (2013). When do dogs help humans? *Applied Animal Behaviour Science, 148*(1–2), 138–149.

Bräuer, J., Stenglein, K., & Amici, F. (2020). Dogs (*Canis familiaris*) and wolves (*Canis lupus*) coordinate with conspecifics in a social dilemma. *Journal of Comparative Psychology, 134*(2), 211–221.

Macpherson, K., & Roberts, W. A. (2006). Do dogs (*Canis familiaris*) seek help in an emergency? *Journal of Comparative Psychology, 120*(2), 113–119.

Marshall-Pescini, S., Schwarz, J. F. L., Kostelnik, I., Viranyi, Z., & Range, F. (2017). Importance of a species' socioecology: Wolves outperform dogs in a conspecific cooperation task. *PNAS, 114*(44), 11793–11798.

Piotti, P., & Kaminski, J. (2016). Do dogs provide information helpfully? *PLoS One, 11*(8), e0159797.

Quervel-Chaumette, M., Dale, R., Marshall-Pescini, S., & Range, F. (2015). Familiarity affects other-regarding preferences in pet dogs. *Scientific Reports, 5*, 18102.

Range, F., Marshall-Pescini, S., Kratz, C., & Virányi, Z. (2019). Wolves lead and dogs follow, but they both cooperate with humans. *Scientific Reports, 9*(1), 3796.

Index

9783030895327